GEOGRAPHICAL STUDIES & JAPAN

GEOGRAPHICAL
STUDIES
&
JAPAN

Edited by
John Sargent &
Richard Wiltshire

Japan Library
Sandgate, Folkestone, Kent

GEOGRAPHICAL STUDIES AND JAPAN

First published 1993 by
JAPAN LIBRARY
Knoll House, 35 The Crescent
Sandgate, Folkestone, Kent CT20 3EE

Transferred to Digital Printing 2004

ISBN 1-873410-15-8 (Case)
      1-873410-19-0 (Paperback)

British Library Cataloguing in Publication Data
A CIP catalogue record for this book
is available from the British Library

Contents

SECTION 2: BRITISH PERSPECTIVES

SECTION 3: THE FIELD EXCURSIONS

Introduction

This short book is not a survey of the geography of Japan, nor is it for the most part an account of how professional geographers - Japanese and others - have interpreted the geography of Japan. Rather, it is meant as a record of a conference, the first of its kind ever held in the U.K., that brought together some 20 Japanese geographers and a roughly equal number of British colleagues for purposes of considering and discussing geographical topics of mutual interest. The conference was held in the summer of 1988 at three successive locations: the School of Oriental and African Studies (University of London), the University of Sheffield, and the University of Durham. At each of these centres, a full day was devoted to the formal presentation of academic papers by Japanese and British geographers, and, since one of the objectives of the exercise was to afford Japanese colleagues with an opportunity to observe at first hand some of the changes currently under way in the English landscape, two days were given over to field excursions - to locations as varied as the London docklands, the Don valley, and the Beamish industrial museum in County Durham.

Following Japanese conventions, the conference was called a "seminar", and indeed the discussions that followed the presentation of papers were often of the lively, informative and amiably contentious kind that one would expect of a good university seminar. English colleagues who attended the seminar learned much about the research vigour and breadth of interest that characterises geography in contemporary Japan; our Japanese colleagues, while sometimes uneasy about the formal aspects of the seminar's proceedings and occasionally a little baffled by the English predilection for concentrating on the past rather than the present, gained valuable insights into the world of professional geography in Britain and observed with interest the changing English landscape, whether via the excursions or from the windows of the long-distance coaches that transported them from one venue to another.

For purposes of providing a complete record of the 1988 seminar, brief descriptions of the excursions are given at the end of the book - these descriptions will perhaps serve to remind our Japanese colleagues of visits to the field that, we hope, were both enjoyable and instructive. But the main business of the seminar, and by far its most valuable component, was the presentation of the academic papers, and it is these papers that make up most of this book. All of the papers contributed by the Japanese members of the seminar are reproduced here in edited form, and the book includes all but two of the papers contributed by English colleagues. We have arranged the papers into two sections, respectively those given by the Japanese and those given by the English participants.

The Japanese papers cover a wide variety of topics (we made no attempt to dictate to participants the subject matter of their contributions) and illustrate the remarkable breadth of interests present within professional geography in present-day Japan. Historical geography in Japan, as in Britain, has undergone a marked revival in recent years, and it is perhaps fitting that the first two papers should deal with historical themes. Given the purposes of the seminar, it is particularly gratifying that each of these contributions links Britain with Japan. Hasegawa compares John Ogilby's "Brittania" road map with Ochikochi Doin's "Atlas of Tokaido" and in doing so skilfully illustrates the practical and ornamental functions of seventeenth-century cartography in both countries. Minamoto, in a paper that discusses the late nineteenth-century foundations from which geography developed as a university discipline in Japan, draws attention, amongst other things, to the little-known role of E.S. Stephenson, an Englishman who taught geography at Tokyo Senmon Gakko, the forerunner of Tokyo University.

Historical studies, whether contributions to historical geography proper or discussions of the evolution of geographical thought, are undoubtedly well-established themes of geographical discourse within Japan. At the same time, though, most of Japan's human geographers are preoccupied with issues pertaining to contemporary urban and industrial growth. This is hardly surprising: the remarkable burgeoning of the Japanese industrial economy since 1955, and the related explosive growth of towns and cities has arguably transformed the Japanese landscape far more dramatically than has been the case elsewhere in the industrial world of the late twentieth century: very few parts of Japan have escaped the influences of industrial and urban growth and the land-use problems that the industrialisation process has brought in its train are virtually ubiquitous. To an extent that is perhaps not fully realised in the West, Japanese geographers for many years have studied a variety of important issues raised by rapid and large-scale industrialisation, and much valuable work has been done on themes such as industrial location (see for example the paper by Morikawa), population migration, regional and urban problems, and regional development policy. In common with other Japanese social scientists, moreover, geographers in Japan have been especially concerned with identifying ways in which the perceived problems caused by industrialisation can best be mitigated: their concern, in short, has always been with the improvement of the human environment and not merely with its theoretical analysis.

One of the main issues that have attracted the attention of Japanese geographers is the domination exerted by Tokyo and its surrounding metropolitan region over the rest of Japan: a domination that is apparent not just in population distribution, but in the geography of the manufacturing and service sectors of the economy. In this volume, Aono surveys the arguments advanced by geographers to explain the concentration of business core functions within the metropolitan region; he also points to the increasingly dominant position enjoyed by Tokyo and its environs as regards the location of advanced research and development facilities. The latter is perhaps a significant trend, for the recent restructuring of the Japanese economy (a process whereby high-technology industries are assuming growing importance in Japan's industrial structure) has done little to weaken the grip of Tokyo over the rest of Japan; indeed the revolutionary development of telecommunications networks, described here by Terasaka, might well, if anything, strengthen rather than diminish Tokyo's metropolitan dominance. Meanwhile, as Terasaka points out, the rise of Japan as a financial superpower has been accompanied by a strong concentration of foreign businesses in Tokyo: Tokyo's emergence as one of the three leading financial centres in the world today is yet a further factor that promises to magnify regional imbalances in economic activity within Japan.

Related to the growth of these regional imbalances is the chronic decline in the relative importance of the primary sector in the Japanese economy: agriculture, forestry and fishing still give employment - or more accurately part-employment - to substantial numbers of people, but their contribution to national wealth has fallen dramatically over the last forty years. Fujita discusses this change as manifested in the predicament of forestry communities - mountain villages that have suffered severe population decline, and whose fortunes, especially in the remoter areas of Japan, have not always been improved by the implementation of official measures designed to alleviate the problems caused by depopulation.

Meanwhile, the growth of towns and cities in Japan has raised many issues of concern to geographers other than questions associated with regional disparities in economic activity. The changing morphology of the Japanese city has provided a major research theme (there exists, for example, a large body of scholarly literature on the central business district and on the Japanese inner city generally), and a substantial amount of research has been done by Japanese geographers on the changing economic functions of the city. Until recently, however, the social geography of the Japanese city has suffered relative neglect, and in this connection the papers by Simko and Ueda are of particular interest. Both contributions consider the role of urban neighbourhood associations in the formulation of local government policy in the Japanese city. Urban neighbourhood associations, which to some extent display the same degree of cohesiveness and sense of identity as rural hamlet communities, have existed in Japan in one form or another for several centuries, and whatever their faults and failings in other directions, one is led to suspect that they have played an important role in accommodating the social strains and stresses associated with rapid urban growth.

Simko and Ueda, perhaps sensibly, avoid such broad speculation. Simko (not a Japanese but a Swiss geographer, whose presence at the seminar gave our proceedings a welcome European dimension) provides a progress report on his research in eastern Tokyo: his objective is to assess the role of voluntary neighbourhood associations in the formulation of policies for the control of atmospheric and water pollution. Ueda is concerned with the mismatch between administrative boundaries - boundaries that have been imposed, as it were, from above - and the more spontaneous territoriality of urban neighbourhood associations. His paper suggests amongst other things that in contemporary Japan, local government initiatives are often doomed to failure unless they adequately take into account the existence - and strength - of local residents' associations.

Although most geographical research in contemporary Japan remains directed towards the investigation of domestic topics, recent years have witnessed a steady growth in studies of the world outside Japan. As is the case in British geography, many of these studies relate to issues of economic development, and are at least in part inspired by a rising concern over the inequalities between the rich and poor countries of the world. At the seminar's Durham meeting, two of the Japanese participants - Koga and Naito - presented papers related to area studies. Both of these scholars belong to the staff of Hitotsubashi University's department of geography, a department that has played an important role in fostering the growth of foreign area studies in Japan.

Koga surveys the development of South Asian studies in Japan, and those involved in area studies in Britain will be painfully familiar with some of the problems that he identifies: the difficulty of communicating across disciplinary boundaries, for example, continues to bedevil area studies in the U.K. as much as it apparently does in Japan. For an example of how geographers can best contribute to area studies, one need look no farther than the paper by Naito. In tracing changes in residential zonation within Damascus, Naito reveals a sure grasp of the exceedingly complicated modern history and politics of Syria. His work, moreover, is informed by a thorough command of relevant French and Arabic-language sources.

Though most of the papers by the Japanese seminar participants are empirical in thrust, it should not be therefore assumed that theoretical work is only a minor concern amongst Japanese geographers. In fact Japanese geographers, perhaps more so than their Western counterparts, have always considered it a methodological imperative to firmly anchor their research work in the appropriate theoretical context. Japanese colleagues, moreover, have made considerable advances in refining Western-derived geographical theory. Two of the papers reproduced in this volume - those by Asami and Kikuchi respectively - are firmly grounded in mathematical theory. Asami illustrates the usefulness of the Box-Jenkins model in measuring local economic fluctuations: his particular concern is to illustrate how the model can be used to analyse employment trends in northern Kyushu. Kikuchi uses an impressive command of mathematical technique to demonstrate how fractal geometry provides a means for greatly refining the central place theory devised by Christaller. Kamozawa's contribution, while far removed from the world of quantitative analysis, also deals with a theoretical issue. His topic concerns theoretical approaches to regional geography, a theme that in the past has exercised the minds of many Japanese colleagues. His argument - that the concept of "consuming space" can bring depth and insight to the analysis of regional problems - may seem novel to some, but is a thought-provoking contention nevertheless.

Even more so than in Britain, Japanese geography shows a sharp dichotomy between physical and human geography, and this division has if anything been perpetuated by the existence of three separate - and sometimes rival - geographical associations, namely the Association of Human Geographers *(Jimbun Chirigakkai)*, the Association of Japanese Geographers *(Nihon Chirigakkai)* and the Association of Economic Geographers *(Keizai Chirigakkai)*. Insofar as the organisers of the seminar on the Japanese side belong to the latter of these three associations, it is perhaps hardly surprising that all but one the Japanese seminar papers ignore the world of physical geography. That said, the contribution of the sole physical geographer on the Japanese side - Hasegawa's paper on the Exe and Nagara river basins - is especially valuable in that it is a comparative study, carefully prepared with the purposes of the seminar in mind. It is to be hoped that if future meetings of the seminar are held, more room will be found to accommodate similarly informative offerings from other of Japan's physical geographers. Japanese work on topics such as Quatenary sea-level changes, climatic history, and prehistoric environments is internationally renowned, and ought to be more strongly represented in the future business of the seminar.

The second part of the book contains papers contributed by British participants in the seminar. One of these, Atkins' paper on Operation Flood, exemplifies the interest of many British geographers in economic development issues. Two others - Sargent's assessment of Japanese studies within British geography and Simmons' paper, which explicitly considers dimensions of Japanese culture as they relate to people-environment relationships - relate to Japan either in whole or in part. The remainder, while not concerned with matters Japanese, nevertheless raise interesting questions concerning the extent to which Western - or British - notions can be successfully applied in the Japanese context. In this connection, the relevance of the theme discussed by White is quite clear: counterurbanization exists in Japan as it does in Western Europe, and although the word "counterurbanization" has yet to be satisfactorily translated into Japanese, there is a growing body of Japanese research work that examines the causes and consequences of the counterurbanization process, including the movement of manufacturing into the countryside (mention of the latter can be found in Fujita's paper).

Johnston's paper, which discusses place and culture, not only emphasises the geographical diversity of cultural attributes in Britain and America, but raises some issues that deserve to be explored in the Japanese context. With one or two regional exceptions (Okinawa is one; Hokkaido, possibly, is another) local cultures of the kind referred to by Johnston are hard to identify in Japan. But the *shitamachi* culture of the eastern inner suburbs of Tokyo (a culture that may have counterparts in other large cities in Japan) represents to some extent a distinctive way of life associated with a definable urban area while in the Japanese countryside, each hamlet tends consciously to preserve its own particular identity and traditions. These and many other Japanese examples can be found to support Johnston's observation that "people are made in places".

These days, place description in Japanese geography tends to be largely a cartographic and statistical exercise: the distinctiveness of place is rarely portrayed by appealing to literary or artistic or any other kind of non-material evidence. Pocock, using Durham as an example, shows that our understanding of the character of a place can be greatly enhanced and deepened by an appreciation of the paintings, poetry, and music associated with the place in question. In this approach, geography becomes more of an art than a science, and an extremely demanding art at that. It is an approach that could well be applied to Japanese situations: Kyoto, Nara, Kamakura, even the Sumida river of Tokyo all spring to mind as examples.

The papers that make up this book, then, show a wide variety of approaches and an extraordinary range of subject matter. We hope that the seminar proceedings convey to readers something of the flavour of what was a stimulating and enjoyable conference; we also hope that the book in its small way will help to raise the consciousness of Japan in British geography.

We should like to extend warm thanks to all of the British participants, who so generously gave of their time and expertise to ensure the success of the 1988 meeting of the seminar. We are particularly appreciative of the assistance of Bryan Coates and Alan Hay (University of Sheffield) and Ian Simmons (University of Durham), and of those of our colleagues who so generously gave of their time to organize and lead the field excursions. We should also like to gratefully acknowledge the invaluable help of Mr David Rose and his colleagues from the Kent Economic Development Board who organized for the seminar a memorable visit to the Chatham Dockyard and the Gillingham Business Park. To all the contributors of papers, Japanese and British alike, we offer sincere thanks, together with profound apologies for the late appearance of these proceedings.

John Sargent,
Richard Wiltshire

Department of Geography,
School of Oriental and African Studies,
University of London

Section 1: Japanese Perspectives

Part I: Historical Geography

Road Atlases in Early Modern Japan and Britain

Koji Hasegawa
Kobe City University of Foreign Studies

British geographers are now recognized as world leaders in establishing new concepts, methods and paradigms in historical geography. Their interest in humanistic geography has in turn stimulated the younger generation of Japanese historical geographers, particularly since the meeting of the International Geographical Congress in Tokyo in 1980. For example, a group was organized in 1981 to conduct research into the Katsuragawa Picture Map, a beautifully coloured manorial map dating from the early 14th century. The group, of which the author was a senior member, classified all pictures in the map into spot, linear and areal elements representing specific landmarks, paths and districts, as suggested by Lynch in his book "The Image of the City". In using this method we tried to capture the spatial perceptions of the map-maker and villagers of that time.

In an earlier paper I discussed language and meaning in Early Modern British city plans: the present paper continues the argument. I will consider two typical road atlases: the Tôkaidô Bungen Ezu (The Measured Atlas of the Tôkaidô Highway, hereafter abbreviated to "Atlas of Tôkaidô"), which is the earliest printed Japanese road atlas from the Edo period, and John Ogilby's "Britannia", the earliest printed road atlas from Restoration Britain. The atlases will be compared from three points of view: first, the map-makers responsible for constructing them; second, the language employed in the atlases; and third, the image of the region during the Early Modern era.

The Makers of the Atlases

Let us begin by comparing the two map-makers, John Ogilby (1600-1676) of "Britannia" and Ochikochi Dôin (c.1627-c.1700) of the "Atlas of Tôkaidô ".

Ogilby, who was born in Scotland in 1600, is famous as an "eleventh-hour map-maker". His early career was spent in the theatre and the book trade, and it was not until after the Great Fire of London in 1666 that he become involved in cartography. Together with his son-in-law, William Morgan, he was appointed to survey the area of the City devastated by the Fire, the result being the publication by Morgan of a large and detailed plan of the City in 1677.

Although Ogilby also published maps of China (1667) and Japan (1669), and surveyed and printed maps of Kent and Middlesex (1672-3), he is probably best known for the first survey of the roads of England and Wales, and for the publication of the results (in 1675) under the title "Britannia, Volume the First: ...". To conduct this survey, Ogilby depended for finance on subscriptions from Charles II and others, the standing lottery in London, and on his salary as the Cosmographer to his Majesty. Shortage of funds subsequently forced him to abandon the original enterprise of publishing "Britannia" in six volumes. Hence we are left with only the first volume, consisting of 100 sheets of road maps and 200 pages of text.

Unlike that of Ogilby, the life of Ochikochi Dôin remains obscure and open to dispute. His name, literally "Far Near Road Print", probably implies that he surveyed and published maps on demand - this would have been his nickname or pen-name. But Dôin is also known to have made the large scale plans of Edo (now Tokyo), published in three editions in 1670, 1676 and 1689.

Professor Yamori of Osaka University has recently identified Dôin as Fujii Hanchi, scion of a feudal clan from Toyama, who as a doctor was familiar both with medical science and with the surveying techniques introduced to Japan from Holland. In 1657, at the age of thirty, Hanchi was dispatched to Edo by his lord, and thus witnessed the Great Fire of Meireki, which occurred in that year. After the Fire the Tokugawa government decided to survey Edo in minute detail, and appointed the Lord of Awa province, Hôjô Ujinaga, to organize the necessary work. Ujinaga too seems to have acquired a knowledge of surveying techniques from the Dutch, in the course of his duties as chief superintendent under the Tokugawa shogunate during 1655-70. Hanchi joined this project together with the chief surveyor, Fukushima Denbei, and other assistants.

Ujinaga had already been asked to survey the principal roads of Japan in 1651, by order of the Shogunate. The "Atlas of Tôkaidô" may be derived from this survey, for Hanchi could have copied the original in the library of the Tokugawa government. The Atlas was published in five volumes in 1690. It is only 26.5cm wide, but has a total length of about 37m, and is folded many times like a concertina screen, a common format in Japanese works of art at that time.

Thus both map-makers were active in the later half of the 17th century, a flourishing age both in economic terms and in the cultural sphere, under the "Merry Monarch" in Britain and the "Genroku Culture" in Japan. And in both capitals, Great Fires left their mark on the development of cartography. Although the detailed surveys of London and Edo had practical purposes, in particular for settling land boundaries accurately after the Great Fires, the techniques of surveying and cartography were soon extended to mapping main roads and to compiling road atlases.

The Language of Maps in the Two Atlases

In general, map language is composed of two categories of aspects: the external aspects, such as form, scale, bearings and ornaments, which together constitute the framework of the map, and the internal aspects, which include pictures, signs and contours.

External Aspects

There is a significant contrast between the two atlases in the actual form that the maps take. The fundamental style of "Britannia" is a strip road map, quite novel in its day. Ogilby describes this style in his preface: it should be projected on an imaginary scroll, with the road starting from a city or town located at the bottom left, then proceeding upwards to end at the top right. As a result, although at first glance the strip maps seem to run parallel to each other, the viewer has to move his eyes in a series of steps. Each strip is about two and a half inches wide and each map has six or seven strips, together forming an imaginary long, vertical scroll, laid out in segments horizontally on the page.

The form of the "Atlas of Tôkaidô" however is much simpler. As already stated, it is a very long folding screen, ranging in length from a maximum of 88cm in Volume 3 to only 66cm in Volume 2. Dôin did not discuss the form in great detail; in his epilogue, though, he did say that he showed the curves of roads as if they were a row of marching ants. This form would have been familiar to Japanese people through the established tradition of picture scrolls.

Hence, the form of the British atlas compels the viewer to move his eyes both vertically and horizontally, while in using the Japanese atlas the viewer's eyes flow more smoothly from the right, Edo, to the left, Kyô (now Kyôto).

As regards map scale, "Britannia" incorporated such innovations as the statute mile and the scale of one inch to one mile (1:63,360), which remained standard in British topographical maps until the 1970s. Ogilby mentions this scale in the third item of his preface, and notes that mileage is to be shown with two dots on the roads, and single dots between mileages for furlongs.

One word in the title of the "Atlas of Tôkaidô" (Bungen) actually means scale, and Dôin explains that the scale is three *bu* (c0.3cm) to one *cho* (c.109m), equivalent to 1:12,000. Of course, the Japanese atlas includes only one trunk road, the Tokaido Highway, so it was easier to adopt a larger scale than Ogilby could for his British atlas.

Both atlases express roads as virtually straight lines, so they have to show bearings from place to place. To each strip Ogilby adds one or two compass roses which show four, eight or sixteen bearings, with a fleur-de-lys for the north point. Dôin, on the other hand, records bearings at each stage town along the Tokaido. The shape of his compass rose is a square in which four, sometimes three bearings are written in Chinese characters: hence bearings between stages are not as accurate as in the British atlas.

Finally, "Britannia" contains many allegorical pictures, especially on its title page. This title was engraved by Hollar and symbolizes the contents of the atlas. Three angels are shown bearing maps of the roads from London to Berwick, the City of London and the County of Yorkshire. Social landscapes of Restoration Britain are depicted on the ground below, including surveyors, map-makers, a grazing scene, a fowler, stage coaches and a windmill. Other pictures depicting social scenes, mythology or even purely geometrical patterns are scattered through the atlas.

By contrast, the "Atlas of Tôkaidô" contains no such pictures. Rather, the map itself is ornamental. The draughtsman of the atlas is Hishikawa Kichibei (alias Moronobu), better known as the founder of the Ukiyoe school of art. Moronobu depicts local customs, intending his pictures to be easily understood by aristocracy and commoners alike. Hence the atlas should really be regarded as the joint product of a map-maker and a painter.

Both atlases contain a variety of pictures, signs and symbols, some of which Ogilby explains in his preface, arranged under ten headings, while Dôin lists twelve items in his introductory remarks. These may be classified as natural or artificial elements and as pictoral or lettered elements. To compare maps between different cultures, however, a more abstract framework would be appropriate. Thus I have categorized the elements in the maps into spot, linear and areal, following the method adopted previously with the Katsuragawa Picture Map.

Linear elements can be discerned in the roads themselves, but the method of expression differs between the two atlases. Ogilby specifies that a road enclosed by a hedge must be depicted by a double black line; otherwise the lines must be pricked or dotted. Turning points are expressed as short double black lines. As the road maps stick faithfully to these, they may be used as historical documents covering the distribution of open fields prior to parliamentary enclosures. The roads run in the centre of each strip with an almost constant width, and the mileage from the milepost at Cornhill, London is given.

In contrast, the Japanese atlas depicts the road in various widths, and almost always lined with trees, the species of which are also depicted. The mileage is not given in figures, but at every *ri* (c3,930m) there are two mounds with trees on them, which are known as *ichirizuka*: distance can be measured by counting these mounds. Turning points are also shown here and there, but not so accurately as in the "Britannia" maps.

The representation of the roads in the British maps appears precise and carefully surveyed, while the Japanese maps seem more conventional and pictorial. In fact, both are based on real measurements: the difference lies in the way these measurements are expressed.

Streams are another characteristic linear element. The British maps have two types of stream: major streams are indicated by three (or more) waved lines and are named, while small streams are shown by one or two lines and are labelled only with the common nouns 'brook' or 'rill'. The "Atlas of Tôkaidô" includes the same types of stream, the smaller ones being depicted by three to five waved lines, and most have their names attached. The treatment of the larger streams is more interesting. A typical example is the River Oi, the crossing of which was one of the most dangerous places along the Tôkaidô Highway. The width of the river has been consciously exaggerated in complete disregard of the scale of the map. People are shown crossing the river, some on horseback, some on foot, and some in a palanquin. These cartographic distortions reflect the contemporary perception of large rivers as dangerous places.

The second group of elements in these maps are the areal ones, including settlements, land use and in particular, mountains and hills. Before the introduction of contours there were various devices for representing relief. In medieval Europe mountains were commonly depicted as "molehills", and Saxton in his "Atlas" (1579) used small cones with shading on one side.

Ogilby introduced a unique way of representing hills, based on the principle that because ascent slopes are drawn as hills in ordinary maps, descent slopes should be shown inverted, or upside down. Thus Ogilby depicted uphill roads by a triangle with the shadow on the right, and downhill roads by a triangle turned upside down with the shadow on the same side. Some indication of gradient was vital in an age of non-mechanised transportation, so Ogilby took care to express gradient accurately and clearly. As a result, hills and slopes in "Britannia" are presented in a variety of ways. For example, while most of the hills are shaded on the right with hatching, some have shading only at the foot, while a few are hatched all over.

The representation of hills on Ogilby's maps is not as metrically accurate as the portrayal of linear features. Rather, distortion has been introduced based on the map-maker's own experience. The real altitude of the hills does not correspond with the size of the triangular figures on the maps, and in some cases relatively gentle slopes are represented visually by the steepest pictorial symbols. Hence, the expression of hills and slopes in "Britannia" is highly picturesque, yet conveys multiple meanings.

Mountains and hills are shown in the "Atlas of Tôkaidô" parallel to the road, as compared with "Britannia", in which slopes are drawn transverse to the roads, and they are drawn from a seaward, or southerly point of view. As a result, correct gradients are not shown whenever the road passes through mountainous regions, not even at Mt. Hakone, which is famous for its steep slopes. In general the mountains are shown pictorially, and are included whether they lie near the road or far from it. As a result, Mt. Fuji appears in the Atlas many times. This landscape technique, typical of Ukiyoe painting, is thought to have been introduced by Moronobu. The Atlas does convey the relative heights of the mountains, even though there is no correct information about the gradient of the roads.

The depiction of settlements has something in common between the two atlases. In Ogilby's map important cities are drawn 'ichnographically' in a plane in accordance with their form and size, while small towns and villages are shown 'scenographically' or in 'prospect' (perspective). Cities are depicted as ground plans, with individual houses shown from a bird's-eye view, and as a result we can discern explicit street plans, and identify these with particular cities. Villages however are aligned lengthwise along the roads, and although their houses too are shown from a bird's-eye view, most villages seem rather stereotyped.

The Japanese atlas also shows village houses arranged in a row along the road, but the street plans of cities are veiled under clouds, even large cities such as Edo, Kyô and Fuchû, with only their castles showing. This reminds us that the primary purpose of the atlas was to delineate the Tôkaido Highway: the Tokugawa government prohibited the publication of detailed city plans.

In contrast to the treatment of cities, villages are drawn in full detail: even the construction materials used in the roofs of the houses are distinguished. In summary then, while the British maps show cities in great detail, it is the villages which receive more detailed treatment in the Japanese maps.

Ogilby's "Britannia" also contains detailed information on land use, through pictorial representation and verbal explanation. For example, parks enclosed by hedges or fences are one of the most characteristic land use features on Ogilby's maps, but because of uncertainty about their size, they are often depicted simply by a semicircle beside the road. Similarly, forests and woods are drawn as groups of trees, each tree casting its own shadow. By contrast, however, arable, pasture and common land are identified only by writing, and their extent and boundaries are not indicated. Nevertheless, if we use the information that Ogilby does give, we can identify nearly all of the land use beside the roads.

In the Japanese atlas, however, there is no equivalent portrayal of land use; indeed, it is not shown at all. Perhaps the land use in those days was almost entirely paddy field, so the map-maker did not think it necessary to show it. Of course, the odd paddy field or lotus pond is marked here and there, but these are ornamental pictures rather than correctly located and mapped landscape features.

Finally, let us consider point signs. Ogilby groups castles, churches and mills with their associated settlements, and had intended to investigate features such as bridges, fountains, mines and antiquities. Dôin too listed places of note and of historical interest, bridges, shrines and temples, and milestone mounds.

The most detailed point element in both atlases is the portrayal of bridges. On the British maps, Ogilby mentions materials, the number of arches and the name, but does not draw the bridge itself. Dôin depicts bridges spanning the rivers and distinguishes small ones from the larger ones, for which lengths are also given.

Among the buildings shown in "Britannia", the churches are the most distinctive. Pictorial signs, with towers and names, appear not only alongside but also far from the roads, implying that churches were thought to be important landmarks for travellers, recognisable from a distance. In contrast, the villages in which most churches would have been situated are not shown. Other buildings shown in detail are mostly specialised facilities, such as mills along rivers or windmills, gallows and beacons on hilltops - all obvious landmarks. They are represented by somewhat stylised pictures, and only wells, indicated by a square with a black circle in it, are shown by the sort of abstract sign characteristic of modern maps.

The Japanese atlas places emphasis on specially selected shrines, temples and beauty spots. In particular, Kyô and Lake Biwa are singled out as important places of note. Local specialities available at each stage are shown, as are carriage charges and the names of relays. In a later pocket edition of the Atlas (1780) a list of carriage charges was added as a supplement.

The most attractive pictures in the "Atlas of Tôkaidô", however, are those of people, ships and meteorological phenomena. There are travellers, peasants and tradesmen in some place, and pictures of pilgrimages or ceremonial journeys made by feudal lords and their vassals in others. Ferries are shown on Lake Hamana and in Ise Bay, while cargo boats plough the sea off Shinagawa and Lake Biwa. Meteorological phenomena include clouds hanging over the large cities or on the mountains. The total effect is very much a landscape in the Ukiyoe style.

In conclusion, although the "Atlas of Tôkaidô" is more pictorial and conventional than the "Britannia", it also represents a synthesis of maps with Ukiyoe art. It rests, moreover, on the solid basis of an actual survey. "Britannia" can be regarded as a transitional stage in the history of English cartography, for although its fundamental character is modern, metric or practical, the conventional and pictorial elements remain, especially in the expression of hills and slopes. Road Atlases serve, of course, as practical guides for travellers, but they also serve as a form of amusement for the masses: hence they have two functions - practical and ornamental.

The Image of the Region in the Atlases

The "Atlas of Tôkaidô" shows a narrow but very long region with Edo as the starting point and Kyô at the other end. "Britannia", on the other hand, shows a sequence of strips starting at the bottom left and ending at the top right of each map. Long distance roads are divided between from two to five maps. The point that needs to be emphasized here, however, is that the starting point in Ogilby's atlas is a fixed one, London. All roads start from London, just as main railway lines and motorways do today. Hence, the viewer of Ogilby's map traces the roads from the capital out to the rest of the country. Of course, it is possible to trace the road in the opposite direction, but pictures and letters are all upside down and it is difficult to read the maps correctly. A map is a mirror of society, and in that sense Ogilby's "Britannia" projects the society of Early Modern England and Wales in a period when the networks of roads, stage coaches and postal services were all being established, as reflected on the strips with roads radiating from London.

By contrast, in the Japanese atlas the Tôkaidô Highway does run from Edo to Kyô, but as the map runs horizontally there is an alternative way of reading it. Moreover, the people on the road are shown travelling in both directions. Although the Tokugawa government promoted a high degree of centralization in the road system of Japan, like the government in Britain, the equivocal relationship between places is peculiar to this atlas.

If we interpret the region in a narrower sense, however, we can see that both atlases convey the image of a region as consisting of a road and its immediate neighbourhood. Generally speaking, roads function as paths, areal elements such as hills, forests and settlements operate as districts, and churches, milestones and other facilities as landmarks. In addition, bridges serve as nodes between roads and rivers, and we can recognize the boundaries of counties on the British maps and the coastline and mountain ridges on the Japanese maps as edges. Road maps are inherently restricted in the way they can express regions to a narrow space. However, the Japanese atlas achieves a depiction of the region as a whole, taking in the entire landscape around the road. On the British maps, the regions give an impression of openness, although they are restricted to the line of the strips.

23

Finally, I would like to add a brief comment on the origin of the form taken by "Britannia". Ogilby himself refers in his preface to the Peutinger map as the best road map available at that time, and of course, the Peutinger map is a form of horizontal picture scroll. However, we can find exactly the same form in the itinerary map for the journey from London to Apulia drawn by M. Paris in the 13th century.

The Beginnings of Modern Geography in Japan: From the Mid-Nineteenth Century to the 1910s

Shokyu Minamoto
Shukutoku College

In this paper I will discuss some of the lesser known figures, both Japanese and Western, who contributed to the introduction of modern geography into Japan. The period treated extends from the mid-nineteenth century, on the eve of the Tokugawa Shogunate's collapse, to the early part of the twentieth century, when academic geography became more or less established at the two leading imperial universities in Tokyo and Kyoto.

I will discuss two categories of people in particular. The first consists of Western experts, engineers and other specialists with geographic knowledge who came to Japan in the nineteenth century, mainly under Japanese government auspices. These "hired foreigners" (*oyatoi gaikokujin*) played an important role in the early stages of Japan's modernization. The second category includes Japanese individuals, most of them independent researchers (as distinct from imperial university faculty or graduates), who helped contribute to the development of modern geography.

Prior to this period, the Tokugawa Shogunate (1603-1867) had adopted a policy of national seclusion, closing Japan's doors almost completely to trade and interchange with other countries. For more than two hundred years, only the small port of Dejima in Nagasaki harbour remained open to intercourse with the Netherlands, China, and Korea.

Under these circumstances, there were two main ways in which the Japanese obtained knowledge of Western sciences - geography among them. One was by reading Chinese translations of Western geography books and maps; the other through study of the Dutch language. In 1720, the Shogunate relaxed its laws to permit the import of Western books other than those relating to Christianity. Subsequently, great advances were made in astronomy and cartography, and some Japanese wrote geographical works based on imported Dutch books.

In the nineteenth century, the advance of the Western powers into the Far East forced Japan to abandon its isolation, and as a result officials in both the Shogunal government and local domains became increasingly interested in world regional geography. This demand for geographical knowledge became more popular and widespread after the Meiji government was established in 1868. The government made "enriching the nation and strengthening its arms" and "increasing production and promoting industry" its main priorities, and part of this drive included the development of mines. Here again there was an urgent need for more detailed geographical knowledge.

Contributions of Western Advisors

Many Westerners worked in Japan during the final days of Tokugawa rule and in the latter decades of the nineteenth century. Investigation of foreign employees is relatively easy in the case of engineers and instructors in geology, mining, surveying and other natural science fields adjacent to geography, for these were practical sciences immediately relevant to the policy of "enriching the nation" and "increasing production". Many experts in mining and geological surveying were invited to Japan from Europe and the United States, among them François Coignet (1837-1902), a French engineer hired by the Satsuma domain (now Kagoshima prefecture) to help develop mines, and Benjamin Smith Lyman (1835-1920), an American who came to Japan to perform fully-fledged geological surveys at the invitation of the new Hokkaido Colonization Office (*Hokkaidô Kaitaku-shi*).

The achievements of these experts impressed upon the Japanese the importance of these fields as realms of academic enquiry, and when Tokyo University was established in 1877, a department of geology and mining was set up within the Faculty of Science. Edmund Naumann (1854-1927), a German, taught in this department.

A list of Western foreign engineers and experts employed by the Meiji government between 1868 and 1889 includes the names of 2,299 persons, including four who taught geography plus nearly forty more who were specialists in geology, mining or surveying.

The following are brief sketches of some of these Westerners who introduced modern geography to Japan.

Willem Johan Connelis Ridder Huyssen van Kattendyke (1816-66)

As foreign ships began to appear in Japanese waters in increasing numbers, the Shogunal government felt the necessity of building a modern Western-style navy, and sought the advice of Jan Hendric Donker Curtius (1813-79), overseer of the Dutch Factory at Dejima. As a result, the shogunate founded a naval academy in Nagasaki in 1855 to provide training for officers. Pels Rijcken (1810-89), a Dutch naval lieutenant, was hired to head the school and thus became the first foreign employee of the Japanese government. Rijcken brought the first Dutch training detachment to Japan, which gave instruction and practical training for a period of about one year, through Japanese interpreters. The curriculum included geography, although exactly what was taught in the geography classes is not known. Among the students at the naval school at that early time was Tsukamoto Meiki (1833-85), who later became one of the chief editors of the government-compiled *Kokoku Chishi* (Regional Geography of Imperial Japan), publication of which commenced in the early Meiji period.

In 1857 a second detachment of training personnel arrived under the command of Sub-Lieutenant Kattendyke. Kattendyke's diary reveals that the curriculum included "Land-en volkenkunde", or "geography and ethnography", which we would probably classify as a form of regional geography in modern terminology. Among the students in the school at that time was Uchida Masao (1842-76), editor of *Yochi shiryaku* (An Outline of World Regional Geography, 1870-80), one of three major best sellers in the early Meiji period.

John Clarence Cutter (?-1910)

Cutter was employed by the Meiji government to lecture on physiology, comparative anatomy and English literature at one of Japan's first agricultural institutions of higher learning, Sapporo Agricultural College. In addition, he served concurrently as medical advisor to the Hokkaido Colonization Office, and taught veterinary medicine and the science of fishery - his lectures on the latter subject are said to have been the first of their kind to be given in Japan. John Cutter remained in Japan from September 1878 until January 1887, longer than most other American instructors, and was very attentive to the health of the students under his care.

Cutter taught geography as part of his lectures on English literature. In 1881 one of his students, Ibuki Sôzô, took copious notes on these lectures, parts of which have recently been translated into Japanese by Ohji Toshiaki (see Ohji, T. (1982): "Sapporo Agricultural College and Modern Japanese Geography." *Jimbun Chiri*, 34, pp. 410-428). The relevant "lecture" is made up of three sections. Section one deals with the geography of Europe, with the first half devoted to a *"regional geographic description"*, in the narrow sense, of Europe and the second half discussing "the origins and evolution of peoples and languages in Eurasia, with a focus on Europe" (*ibid.*, pp. 423-4, my emphasis). Section two covers the geography, history and languages of Great Britain, and the third section treats the history of English literature, with an emphasis on Shakespeare. As Ohji points out, Cutter introduced into Japan the geographical concepts widely taught elsewhere in the nineteenth century, such as the idea that the ratio of coastline to land area determined the degree of civilizational development in a region.

As we shall see later, Uchimura Kanzô and Shiga Shigetaka were prominent figures in Japanese intellectual history who made great contriutions to the spread of modern geography in Japan outside of academia. Both studied at Sapporo Agricultural College, Uchimura from 1877 to 1881 and Shiga from 1880 to 1884, but we do not know whether or not they attended Cutter's lectures on the geography of Europe.

Charles Mettauer Bradbury (1862-?) and Edward Standley Stephenson (1871-?)
Little research has been done hitherto on Westerners employed by private Japanese organizations in the early days of Japan's modernization. Here I shall mention two such people, both of whom lectured on geography at the Tokyo Senmon Gakkô, an institution founded in 1882 which was later to become Waseda University.

The first of the two is C.M. Bradbury. His personal history, typewritten in English and signed by Bradbury himself, is preserved at Waseda University's Office of University History. It states that Bradbury studied chemistry, geology and mineralogy at the University of Virginia in the United States, graduating in 1885, and obtained his Ph.D. from the same university three years later. He travelled to Japan in September of 1888 and taught "English (reading and conversation)" and "Asian geography" at Tokyo Senmon Gakkô from September 1898 until around 1901. Contemporary documents tell us that "Asian Geography (English Text)" was a subject in the school's higher preparatory course at that time. It is not known, however, what text was used or what specifically was taught.

The other Westerner I would like to discuss is E.S. Stephenson. His personal history, also preserved at the Office of University History, states that he was born in the County of Cheshire in 1871 and graduated from King William's College on the Isle of Man and from Dulwich College, London. He completed studies in geography at both institutions and came to Japan in 1897. At Tokyo Senmon Gakkô he taught

English (composition and conversation) and geography from December 1901 to July 1903. Contemporary records from 1903 show that geography was taught in the school's higher preparatory course, under the titles "Geography (Western)" and "Geography and [English] Conversation (English texts published by this school [Tokyo Senmon Gakkô])." Stephenson's personal history contains the note "at present using Clarke's Geography." "Clarke's Geography" probably refers to the *Class Book of Geography* (1st edition, 1878) by C.B. Clarke. The 1901 Tokyo edition of this text is in the collection of the Waseda University Library, with the inscription on its title page, "Presented by the University Press [of Tokyo Senmon Gakkô] on October 3, the 34th Year of Meiji [1901]." This fact, and a passage in Stephenson's curriculum vitae noting "as a basis for conversation - that is merely paraphrasing and explaining the words of C.B. Clarke F.R.S.," imply that Stephenson used the *Class Book of Geography* in his "Geography and Conversation" class as an aid to English study. The *Class Book of Geography* is devoted mainly to the regional geography of various parts of the world, and I would judge that in comparison with Japanese geography texts of the time, it was on a par with textbooks in use at teachers colleges and secondary schools.

It was at this time, incidentally, that Shiga Shigetaka was teaching "Western political geography" in the higher preparatory course at Tokyo Senmon Gakkô. Further work needs to be done on Bradbury and Stephenson and the content of their teaching, and I hope to publish the results of my research on this subject in the near future.

Japanese Non-Academic Geographers

Prior to the establishment of academic geography in Japan around 1910, a small number of Japanese had authored geographies under the influence of modern Western geography and/or works on contemporary Japan based on the principles of modern geography. Several translations and original works in physiography and physical geography were published by academic specialists in geology such as Fujitani Takao and Kôtô Bunjirô (both of Tokyo University). In human geography, however, it was individuals outside academia who made the greatest contribution.

Shiga Shigetaka (1863-1927)
Shiga was born into a samurai family in the Province of Mikawa (now Aichi Prefecture). In July 1884 he graduated from Sapporo Agricultural College with a bachelor's degree in agriculture. He was not only a well-known writer on popular geography but also an influential thinker and opinion leader. He was the first to apply

the "comparative method" advocated by the English geographer and educator J.M.D. Meiklejohn (1830-1902) to Japan. Underlying his geographical works was a firm belief in a brand of nationalism that was compatible with internationalism. His basic and consistent concern was with Japan's place in the international community. In one of his major works, *Nihon fukeiron* (Japanese Landscapes, 1894), he vividly described the beautiful landscape of Japan and tried to make his countrymen conscious of Japan's position in a world context. In his later years he wrote *Shirarezaru kuniguni* (Unfamiliar Countries, 1925), as an introduction to the countries of the Middle and Near East, (an area in which his contemporaries showed almost no interest), in which he emphasized the importance of the region in international politics.

Shiga's geography dealt with more than mere geographical phenomena, however. As an enlightened thinker, he adapted the principles of Western geography to create and popularize a "Japanese Geography."

Uchimura Kanzô (1861-1930)
Uchimura, too, attended the Sapporo Agricultural College, graduating in 1881. He was converted to Christianity while a student, and in 1885-86 studied at Amherst College in Massachusetts. He was a pious Christian, known internationally. Though they were both Sapporo Agricultural College graduates, Shiga and Uchimura differed greatly in their attitudes toward religion: both were important figures in the intellectual history of modern Japan.

Uchimura wrote fewer geographies than Shiga, and here I shall focus upon one of them, his *Chirigaku-ko* (Discourse on Geography, 1884: the title was changed in 1887 to *Chijin-ron*: Discourse on the Earth and Man). As a preface to the main text of this work, Uchimura gave a list of more than ten foreign works (chiefly geographies), including *The Earth and Man* (1849) by Arnold Guiyot (1807-84), noting them as his direct and indirect sources. *Discourse* mainly discusses the Asian, European, American and other continents, and what characterizes his thought in this work is the idea of the "mission of Japan". He stressed how important it was for the Japanese people to know the distinctive geographical features of their country and realize its mission: namely, to act as a mediator between East and West. In that sense, *Discourse* resembles the popular geographies of Shiga.

Other important non-academic geographers include Yazu Masanaga (1863-1922), author of *Nihon Chimongaku* (Physiography of Japan) (1889), a discussion of geographical phenomena in Japan based on modern geography, and Makiguchi

Tsunesaburô (1871-1944), author of *Jinsei Chirigaku* (Geography of Human Lives, 1903).

As academic geography became established, these pioneer geographers faded into obscurity. Their ideas on human geography deserve to be reexamined, however, in the context of a Japan suffering from increasing damage to the environment brought by rapid economic growth and industrialization.

Part II: Economic Geography

Research into the Location of Business Core Functions in Japan

Toshihiko Aono

Chuo University

Only since the late 1960s has much attention been paid in Japan to the location of business core functions and to their role within the regional structure of the national economy. The rapid expansion of heavy industry from the mid 1950s to the early 1970s led inevitably to the growth of large corporations within key industrial sectors and the formation of an oligopolistic national economy. As these corporations grew, their decision-making and administrative functions became ever more crucial to survival and expansion. Each company established administrative sections responsible for the execution of these functions, while increasing "oligopolistic interdependence" necessitated enhanced facilities for collecting external information.

Meanwhile, the growth of heavy industry caused a number of regional problems which needed to be solved quickly, for both economic and political reasons, and increasing regional disparities between congested cities like Tokyo and the rest of the country focused particular attention on the concentration of business core functions in the Tokyo Metropolitan Area (MA).

Most early observers viewed the increasing agglomeration of business core functions in large metropolitan areas as being economically rational, and argued that core functions and service industries should be encouraged to develop in metropolitan areas, while manufacturing activities should be relocated elsewhere.

At the same time, however, a small number of radical researchers interpreted this phenomenon in terms of the growing spatial division of labour created by monopolistic capital. They suggested that the concentration of large companies' headquarters in Central Tokyo could be attributed primarily to the close relationship between large companies and government bodies. Meanwhile, another group of researchers focussed on changes in the nationwide hierarchical urban system, which reflected the degree of agglomeration of head or branch offices in each area.

Since the first oil crisis (1973), the growth of the Japanese economy has slowed down, and the emphasis of industrial policy has shifted from traditional "smoke-stack" industries to high technology "knowledge-intensive industries".

Internationalization and the emergence of an information-oriented society have made information-collecting functions even more necessary for corporations, and research and development (R&D) functions are expected to play a decisive role in competition between rivals both at home and overseas.

As a result, the location of business core functions has attracted widespread attention in the 1980s, particularly with the publication of two significant government regional policies: the Technopolis Plan and the Fourth Comprehensive National Development Plan (CNDP). The Technopolis Plan aims to promote regional development through the construction of Technopolises based on high technology industries, R&D facilities (including universities), and superior residential environments. The creation within or attraction of R&D facilities to the eighteen areas designated so far is one of the cornerstones of the Plan. The Fourth Comprehensive National Development Plan, approved in July 1987, aims to correct the concentration of high-level urban functions and population in the Tokyo MA, and to promote the balanced development of regions. For example, in July 1988 the government decided to relocate 79 governmental bodies from the capital area as part of its decentralization policy, in line with the Fourth CNDP. In the 1980s, then, increasing emphasis has been placed on the contribution that core functions and high technology industries are expected to make to the development and stabilization of the regions. This in turn has stimulated much greater academic interest in this subject.

The high degree of concentration of economic activities in the Tokyo MA is illustrated by the following figures, taken from the 1988 Economic White Paper, on the percentage shares held by this area out of the appropriate national totals in 1986:

Area	*4 %*
Population	*25 %*
Number of business establishments	*23 %*
Annual manufacturing output	*25 %*
Annual commercial sales	*39 %*
Value of bank loans	*55 %*

The concentration of corporate headquarters is even more marked. For example, of a total of 945 large corporations listed on the First Section of the Tokyo Stock Exchange (TSE) in 1983, 57.1 percent had their head offices in Metropolitan Tokyo, and another 16.6 percent in Osaka Prefecture. In 1983 Metropolitan Tokyo also contained 69.9 percent of the headquarters of the 186 major member companies within the six largest "groupings" of companies *(keiretsu)*, while 59.0 percent of the 1,031 large manufacturing corporations listed on either the First or Second Sections of the TSE in 1982 had their head offices within the ward area of Tokyo. Moreover, most of these corporate headquarters are located within the three wards of Central Tokyo.

However, the concentration of R&D establishments presents a slightly different picture. In 1986 the three largest MAs (Tokyo, Osaka and Nagoya) together accounted for more than 70 percent of the 835 free standing R&D facilities operated by private manufacturing concerns: the Tokyo MA alone had nearly half the total. These R&D facilities tend to be located in the outer suburbs of the MAs, however, rather than in the cores. Kanagawa prefecture for example supports a concentration of R&D facilities which almost rivals that within Metropolitan Tokyo.

There is widespread agreement amongst Japanese geographers and economists that the concentration of business head offices in large metropolitan areas is mainly attributable to "agglomeration economies" or "external economies" derived from the ease of procuring specialized information. "Convenience for procuring business information from other companies and industrial circles" is the most important reason cited by companies for having their headquarters in the Tokyo area. Hence the main locational factors in Japan seem to be little different from those in other advanced countries.

"Convenience for liaising with central government bodies", which are heavily concentrated in Central Tokyo, is the second most important locational factor, and the larger the capital endowment of a company, the more crucial is contact with the appropriate government agencies. To some extent this reflects the close relationship between business circles and conservative political and bureaucratic groups in Japan, but it also mirrors the similar close relationship widely observed in other advanced nations which have strong centralized political systems, such as the United Kingdom and France, where business core functions also tend to be concentrated in the capital area.

Another important locational consideration for the headquarters of large corporations is the "financial convenience" factor. Large Japanese corporations raise investment funds mainly through borrowing from large banks and insurance concerns, whose headquarters are also highly concentrated in Central Tokyo. Although the concentration of financial establishments in the largest metropolitan area is paralleled in other advanced countries, it is presumably less important as a location factor in other countries because large corporations there tend to depend more on their own accumulated profits as the major source of investment funds.

Some commentators have suggested that the introduction of information technology has reduced the importance of direct personal contacts for the execution of headquarters functions in countries such as the United Kingdom. In Japan, however, the importance of face-to-face contacts seems to have remained unchanged, though why this should be so has yet to be explained.

According to the 1988 Economic White Paper, economic activities are becoming ever more concentrated in Metropolitan Tokyo chiefly because of the expanded business opportunities created by the internationalization of the national economy and innovations in information technology. It also points out that of all economic activities, it is business core functions in particular that have become concentrated in the Tokyo area. This trend has been confirmed not only by government economists but also by academic researchers.

Over the 1965-87 period the relative concentration in Metropolitan Tokyo of the head offices of leading companies listed on the First Section of the TSE changed very little, from 58.7 percent of the national total to 58.4 percent. Two qualifications are necessary though. First, during this period the number of companies listed on the TSE increased by 77.5 percent, and some 58.1 percent of this increase took place in Metropolitan Tokyo. Second, during this period many provincial banks which had their head offices outside the Tokyo area joined the TSE. When financial corporations are excluded from the figures, Tokyo's concentration ratio increases from 58.7 percent in 1965 to 59.3 percent in 1983, and further to 60.7 percent in 1987. As far as the financial sector is concerned, therefore, Tokyo's core function has been strengthened.

This continuing concentration of business headquarters functions in the Tokyo area can also be illustrated in other ways. For example, foreign-owned subsidiaries have rapidly increased in number in recent years, and 86 percent of them have chosen to locate their headquarters in the Tokyo area. As the Tokyo financial market has become increasingly internationalized, so many foreign financial institutions have opened offices in the Tokyo Central Business District. The number of foreign banks who have their own offices in Metropolitan Tokyo increased from 166 in 1980 to 204 in 1986, while the number of foreign securities concerns rose from 68 to 161 over the same period.

But is this tendency towards increasing concentration a distinctive characteristic of the regional structure of the Japanese economy, or one that is shared with other advanced countries? The trend in the U.S.A. certainly seems to differ dramatically from that in Japan. In America the concentration of large company headquarters has shifted over time from the North-East, through the Mid-West, to the South and West as new industries have developed. In Britain, on the other hand, the regional concentration of headquarters functions appears to be similar to that in Japan. According to Goddard and Smith ('Changes in corporate control in the British urban system', *Environ. Plann. A*, 10, 1978), Greater London and the rest of the South-East recorded net gains of 19 and 17 headquarters respectively between 1972 and 1978 from among the leading 1000 companies selected. The important difference, however, is that although South-East England as a whole enjoyed rapid growth in the number of headquarters, Greater London experienced a very low rate of increase. This trend seems to have continued thereafter, given that many of the newer high technology companies have located their head offices outside Greater London, and particularly in the rest of the South-East. Increasingly, foreign-owned companies have chosen their head office sites in the rest of the South-East beyond Greater London as well.

Hence, we must conclude that the *growing* concentration of large corporate headquarters in Metropolitan Tokyo, and especially in Central Tokyo, is indeed a distinctive characteristic of the regional structure of the Japanese economy.

The high *level* of regional concentration of corporate headquarters, as opposed to the concentration *trend* over time, is also often regarded as a distinctive characteristic of the regional structure of the Japanese economy. This, however, seems less likely. For example, an international comparison of the degree of metropolitan concentration of corporate headquarters makes it clear that Tokyo's concentration ratio lies in the middle range for leading metropolitan areas in the advanced countries. London has a higher concentration than Tokyo, as does Paris. In some centralized unitary states such as France, the United Kingdom, Japan and Sweden, headquarters are highly concentrated in the capital area, while in some decentralized federal states such as the U.S.A., West Germany and Switzerland head offices are much less concentrated in the leading metropolitan area. Yet the concentration of headquarters in Tokyo is not particularly high even in comparison with other countries in the first of these two groups.

One thing that is distinctive about Tokyo, however, is the extent to which its dominance extends across all sectors of industry. In all industrial sectors except one (electicity and gas supply), Tokyo's concentration ratios are higher than those for all other prefectures, and the same is true for all fifteen classes of activity within the manufacturing sector. In the U.S.A., however, the location of large corporate headquarters is much less concentrated within each manufacturing sector and in manufacturing industry as a whole. Amongst the seventeen major classes of activity in the American manufacturing sector only one, rubber and tires, has more than half of its large corporate headquarters within one state, Ohio, and while New York State holds the largest number of corporate headquarters in seven out of the seventeen sectors, the first places in the remaining ten are occupied by eight different states.

In Japan as in Britain, high technology industries and R&D activities have been hailed as an effective means of revitalizing regional economies: this is the basis, after all, of the Technopolis Plan. Yet as we have seen, the location pattern of R&D establishments in Japan is characterized by strong concentration within the three large metropolitan areas. The regional R&D concentration ratio for the Tokyo MA alone (49.3 percent in 1987) is roughly equivalent to that for the South-East region of Britain (45.5 percent in 1983). In contrast, R&D activities in the U.S.A. are characterized by a far more dispersed location pattern. Japan and Britain also share an identical locational trend for R&D activities. In the period between 1982 and 1985 Japan experienced a "second rash" of R&D facility construction, and during the peak years the concentration of R&D activities in the Tokyo MA strengthened considerably. Contrary to all expectations then, it seems unlikely that large companies will choose sites outside the Tokyo MA for their affiliated R&D establishments. Within the Tokyo MA, however, the concentration ratio for Metropolitan Tokyo itself has decreased, while the ratios for the four neighbouring prefectures (including Ibaraki) have increased. Kanagawa secured the largest share of new R&D facilities established between April 1983 and September 1985, and is expected to overtake Tokyo in the very near future. This trend closely matches that reported by a number of writers in the United Kingdom.

In both Japan and the United Kingdom, the ease of procuring specialized information is considered to be the most critical factor in the location of R&D establishments. Certainly it is this factor that has influenced the location of R&D establishments in the South-East of England and in the Tokyo MA. The most significant difference between analyses of the determinants of R&D location in Britain and Japan concerns the role of the government and its R&D establishments. British researchers generally stress this role, while Japanese researchers either neglect it or place little emphasis upon it. This in turn reflects differences between the two countries in the role of government in relation to the private sector's R&D activities. In the United Kingdom, military procurement has an important influence on the location of much private sector R&D activity, to the extent that it is possible to detect a "military-industrial complex" in the R&D field. In Japan, on the other hand, industrial R&D activities have been heavily dependent on private companies. In 1983 68.2 percent of Japan's total R&D expenditure was accounted for by the private sector, and this figure has since risen further.

Another difference between the two countries concerns the presence of sufficient numbers of scientific and technological workers and of superior residential environments to attract and retain them. British researchers tend to emphasize this factor not only in the location of R&D activities but also in the location of high technology industries and in the industrial "urban-rural shift." Although Japanese researchers do refer to this factor, few claim that it has actually influenced the location of R&D facilities: at most, this factor tends to be a desirable condition for attracting R&D activities rather than a locational determinant.

Some researchers claim that corporate headquarters and R&D functions have become crucial not only for the growth and survival of individual companies, but also for the stability and prosperity of regional economies. As far as Britain and Japan are concerned, however, the reality is that the concentration of these functions within several relatively small areas has intensified. It is clear that not all regions can expect to attract these functions: hence each region should determine the most feasible way of revitalizing its economy based on the actual and specific conditions each confronts.

The Transformation of Regional Systems in an Information-Oriented Society

Akinobu Terasaka
Ryutsu Keizai University

The evolution of new media in contemporary Japan

In this report, I will consider some of the spatial problems arising from the transmission of information in Japan's increasingly information-oriented society, from a geographical viewpoint. The term information has gradually acquired a new depth of meaning. The rapid transformation of Japan into an information-oriented society has been responsible, along with the development of the enabling technology, for the formation of new information networks. These networks have caused a diminishing of time and space and a restructuring of spatial organization, a transformation which has in turn had impacts upon regional systems.

Mass media methods of transmitting information, including newspapers, magazines, radio and television, have long been familiar features of everyday life, as have certain types of personal communications media such as the telephone and the mail. New media, on the other hand, is a term referring to new methods of disseminating information, such as the computerized disposition of data, telecommunications, satellite transmission and so on. These methods evolved from earlier developments, such as the automation of telephones and the rapid diffusion of telephones and television into private homes during the 1960s, which in turn reflected the high rate of economic growth at that time.

The new media have now permeated society, both at the industrial and at the individual level. In the year 1972, with the introduction of the first phase of the government's liberalization of telecommunications circuits, a public circuit was made available in Japan for general use, together with the necessary data processing systems. This made possible the mass utilization, rapid dissemination and regional diffusion of large volumes of information. Specific examples of these developments include the establishment in 1964 of the "green window", an on-line system for ticket purchases and reservations, by Japan National Railways (now Japan Railways, following privatisation), using its own network, and the mechanization of banks based upon the introduction of CD (cash dispenser) systems from 1969 and ATM (automated teller machine) systems from 1971. In this way, the linking of computer systems with telecommunications circuits became an inseparable part of the daily life of ordinary people.

The second phase in the liberalization of telecommunications circuits took place in 1982, and at the same time the emergence of digital systems opened up new fields in communications.

From September 1984 to March 1987, the Nippon Telegraph and Telephone Corporation (NTT, another privatised company) carried out an experiment involving a model INS (Information Network System) in the Mitaka and Musashino areas of Metropolitan Tokyo, typical suburban residential areas located about 20 km west of the city center. New machines with digital communication capabilities and equipped with optical fiber cables, enabling high speed transmission of visual patterns, were installed in industrial enterprises, municipal offices and private homes. Some 2,000 households participated in these experiments, with the information being processed through around 300 enterprises.

The model system consisted of two separate networks; a 64 kb/s digital communications network and a regional communications network for transmitting visual patterns. These were connected to existing communications networks with the purpose of experimenting in new types of services, such as digital telephones, visual communications (telephone calls and televisual conferencing), facsimile communications, video-tech communications and data processing.

NTT has been working on a new ISDN (integrated services digital network), which has offered a wide range of commercial services since April 1988. This project is a successor to the INS model system mentioned above. Initially, service is being provided in the Tokyo, Osaka and Nagoya metropolitan areas. By the end of 1988, however, the ISDN network will have spread to fourteen other major cities.

It is clear that a great deal of thought still has to be given to a number of outstanding questions, including the methods to be adopted by national and local governments, NTT and other enterprises to deal with the new media, including the costs involved, all of which have yet to undergo any form of institutionalization. From the technological standpoint, there are still problems to be resolved in order to adjust the transmission systems adopted in this experiment to the worldwide standard. While the experimental model has proved to be potentially effective in bringing about positive changes at the individual, corporate and regional levels, it has also become obvious that a considerable length of time, as well as enormous sums of money, will be required in order to bring about the expansion of these new types of services at the nationwide level.

Concentration of central administrative functions necessitated by access to information

So far as the location of enterprises is concerned, the rapid growth of information diffusion techniques has reduced the resistance to communication over long distances from the center, and it was once thought that this would exert some control over the concentration of enterprises in large cities. On the contrary, however, in recent years there has been a sharp increase in the location of offices in Tokyo proper (the twenty-three wards of Tokyo), and especially within the three wards of Inner Tokyo (Chiyoda, Chûo and Minato). In 1969, 56.5 per cent of the head offices of enterprises listed on the First Section of the Stock Exchange were located in Tokyo, while 20.3 per cent were located in Osaka, but by 1981 the percentage of head offices located in Tokyo had risen to 62.3 per cent, while the proportion in Osaka had fallen to 13.5 per cent. Moreover, the 1981 Establishment Census showed that some 23.1 per cent of all offices in Japan were located in Tokyo, but if we take into account the capitalization of the enterprise concerned, we find that the larger the scale of the enterprise, the more likely it is that it would be located in Tokyo. Hence 56.8 per cent of enterprises with capital of over five billion yen were located in Tokyo, while for enterprises having capital of between 1 billion - 5 billion yen and 100 million - 1 billion yen, the shares were 52.6 per cent and 38.2 per cent respectively.

Amidst this rapid transformation of Japan into an information-oriented society, Tokyo has also been consolidating its function as an international financial market. As a result, an increasing number of foreign enterprises are at present busily locating their offices in Tokyo. According to Ministry of Finance sources, in 1974 some 279 foreign establishments opened new offices located in Tokyo (of which 150 were non-manufacturing companies). Thereafter the number increased year by year until, in 1982, it reached 412 (of which 292 were non-manufacturing companies), and in 1984 490 (of which 340 were non-manufacturing). The majority of these offices were located in Inner Tokyo: in 1984, of the 2,256 foreign companies in Japan, 1,406 (62.3%) were located in Inner Tokyo proper, while only 353 (15.6%) were located in areas other than Tokyo.

Thus, it is clear that an overwhelmingly large concentration of foreign establishments is to be found in Tokyo, the concentration of foreign financial organizations in the center of Tokyo being particularly marked. Along with the rapid transformation of Japan into an information-oriented society, the current internationalization of Tokyo businesses, for which this transformation is a prerequisite, will further intensify the concentration of offices in Tokyo. As a result, CBD expansion is currently taking place. Moreover, the urban structure of Tokyo further reinforces the advantages of concentrating central management functions in the inner areas of the city.

Conclusion

While a number of problems have yet to be resolved before the new media systems, such as INS and urban type CATV, become fully workable, the Ministry of Posts and Telecommunications' Teletopia Project and the Ministry of International Trade and Industry's New Media Community Concept both have the potential of inducing enormous transformations in local communities and in the lives of the people at large. In addition, the rapid development of information service industries will exert a powerful influence on the industrial structure as a whole.

A Survey of Recent Geographical Studies of Manufacturing in Japan

Shigeru Morikawa
Osaka University of Economics

Studies of manufacturing by geographers in Japan began in the 1930s, since when a close link has existed between the development of research and the progress of industrialization. In particular, there was a marked upsurge in interest during the period of rapid economic growth between 1960 and 1970. During this period, geographers focussed their attention on locational changes in manufacturing and on the formation of new industrial regions, both of which had resulted from the rapid development of manufacturing coupled with the rise of heavy industries (such as iron and steel, oil refining, petrochemicals, general and electrical machinery, automobiles etc.) in place of light industries (especially textiles). Since the Oil Crisis of 1973, however, geographical research on manufacturing has tailed off markedly. The switch to a low growth economy, together with the relative lowering of the position of manufacturing and the rise of the service industries in their place, caused geographers to become less interested in manufacturing and to turn their attention towards tertiary industries and urban geography instead. This trend has continued throughout the 1980s - the subject of the present report.

Geographical studies of manufacturing in Japan can be divided into four general fields on the basis of their contents and methods of study. The first field consists of empirical studies dealing with the location of individual types of manufacturing at the national level and with the production system of individual industries on a regional scale. Typical examples include the study of the location of the iron and steel industry in Japan and of the production system and spatial distribution of the machinery industry in the Tokyo Metropolitan Region. The second field covers studies that deal with the spatial organization of manufacturing in Japan, the development of manufacturing within a particular region, and the structure of manufacturing in industrial areas. The third field comprises research on the industrial location policies and regional policies of the central government, such as the Comprehensive National Land Development Plan and the New Industrial Cities plan (*Shin Sangyô Toshi* in Japanese). Finally, the fourth field includes work within the realm of industrial location theory, such as discussions of Weber's industrial location theory. The characteristics of each of these fields of study in the 1980s will now be examined in detail.

The first field, locational studies dealing with specific industries, has long represented the major focus of Japanese manufacturing geography, a position which did not change during the 1980s. The main foci of research during this period were as follows:

a) Studies of the locational reorganization of industries such as iron and steel, shipbuilding, oil refining and petrochemicals, industries which had once been the growth sectors of the Japanese economy. Particular attention was devoted to the plant closures and production cutbacks which took place in these industries in the early 1980s. Related studies have dealt with the influence of locational reorganization on economic activity in company towns (such as Kamaishi and Aioi).

b) Empirical studies of the location of high-technology industries, which have undergone rapid development in recent years. High-technology industries have tended to agglomerate in Japan's metropolitan regions (and especially in the Tokyo Metropolitan Region), because of the great importance attached to research and development functions as locational conditions. On the other hand, the integrated circuit (IC) industry, which does not rely upon direct access to research and development, has become located in more provincial areas (such as Kyushu), because of the importance of labour supplies and access to airports. This locational tendency is reflected in the content of these studies, which deals with the relationship between the location of high technology industry and airports and with high technology industry's influence on other industries in the locality.

c) Studies of the production systems (the structure of employment and sub-contracting systems) in the general and electric machinery industries, both of which experienced a degree of dispersal to more provincial areas from the metropolitan regions during the period of rapid economic growth.

d) Studies of community-based small and medium sized industries (*jiba sangyô* in Japanese) located in particular towns and villages. Research of this kind has been one of the mainstreams of manufacturing geography in Japan and consequently the number of studies of *jiba sangyô* has been rather large as compared with other work undertaken during this period. These studies deal mainly with the historical development and current operating conditions of these industries in particular localities. Other studies have investigated the role that *jiba sangyô* can play in revitalizing the regional economy of those

44

localities which have been severely affected by the reorganization of Japan's industrial structure.

Typical studies within the second field were concerned with the spatial reorganization of manufacturing on the national level. These studies made it clear that the transformation of the industrial structure during the period of low economic growth had brought about changes in the pattern of spatial organization which had been inherited from the period of rapid economic growth. Other studies within this field concentrated on small and medium sized enterprises, which account for the greater part of manufacturing activity in the metropolitan regions. These studies focussed in particular on the relationship between small and medium sized enterprises and the community, the role of small and medium sized enterprises as incubators of technological innovations, and the relationship between small and medium sized enterprises and the problems of the inner city.

One of the characteristic areas of research within the third field dealt with the evaluation and review of the regional policies and industrial location policies which had been adopted by the Japanese government. Studies of this type focussed in particular upon the present health and future prospects of industrial activity in local areas, since the first objective of both sets of policies was the promotion of industrialization within local areas through the dispersal of industries from metropolitan regions. Attempts were also made to analyse the formation of industrial estates which had been constructed with the intention of attracting small and medium sized enterprises. Other work covered the high-technology industry area plan (the *Technopolis* programme) which was developed by the Ministry of International Trade and Industry as a means of promoting the transformation of the nation's industrial structure. The objective of this plan is to establish institutions of research and development and to locate high technology industries in local areas, with the ultimate aim of promoting the development of those areas. At present, twenty areas have been designated under this plan throughout the country. Studies which have focussed on this plan have described the present situation in the areas affected and assessed the proposals which have been put forward to realize the plan.

As for the fourth field, the theoretical approach to the location of manufacturing activity has always been somewhat neglected in Japanese geography, a tendency which continued through the 1980s. Moreover, those studies which were undertaken in this period were mainly reviews of existing industrial location theories (Weber's theory, Hoover's theory etc.). In addition, essays were published on the theory of industrial location and regional structure, chiefly by Marxist geographers.

In conclusion then, as this review has made clear, a wide variety of geographical studies of manufacturing activity were published in the 1980s, largely in connection with the transformation of Japan's industrial structure which took place during this period. It is to be regretted, however, that no new methods for the geographical study of manufacturing were put forward. Most of the studies undertaken during this period were empirical, and it must be said that they have not led to the development of any new theory.

Changes in Mountain Villages and Policies for the Development of Mountain Areas in Japan

Yoshihisa Fujita
Aichi University

The purpose of this paper is to examine the changes that have taken place in the mountain villages areas of Japan, and in particular those changes which have occurred since the period of rapid economic growth in the 1960s.

Geographical conditions of mountain villages

About 70% of Japan's land area is mountainous and these mountain areas are divided into two parts by the Fossa Magna - the fault corridor that traverses central Honshu from the Pacific to the Sea of Japan. To the northeast of this line the density of mountain villages is lower because of the cooler climate. The area to the southwest is further divided into two zones by the median tectonic line (a well-defined line of faults that runs from the Japan Alps in the east to Kyushu in the west). Within the "outer zone", which lies south of the median line along the Pacific coast, a combination of depositional strata and heavy rainfall has created steep physical features, with a high proportion of settlements located on valley slopes. The "inner zone" to the north is largely underlain by granite: hence the topography is more subdued, consisting mainly of basins and plateaus. Settlements here tend to be located at the foot of gently-sloping mountains.

At present the population of these mountain areas is less than ten million, or between seven and eight percent of the total population of Japan, and about 15% of the population consists of people who are already more than 65 years old.

The formation of depopulated and low income areas

From about 1960 onwards, large numbers of people began to stream into Japan's metropolitan regions from mountain areas and remote rural localities, under the stimulus of new government policies to boost national income. About 30 million people moved to urban areas in the two decades after 1960, a scale of migration unprecedented in Japan's history. In the process mountain areas lost much of their population, and severely depopulated areas emerged all over the country. The proportion of older people in these areas increased sharply, many villages and hamlets vanished, and more than 100 settlements had to be relocated to more convenient sites.

These phenomena came as a surprise to many Japanese scholars, who up to this point had tended to stress the stability and inertia of rural settlements in Japan. In their subsequent efforts to understand the new depopulation trend, however, most of these people failed to conduct detailed studies, preferring instead to base their conclusions on the casual observation of conditions in out-of-the-way places. Most of these scholars were city people labouring under a disadvantage in attempting to study places that were essentially foreign to them. Nevertheless, their results were adopted by government planners as the basis for regional policies for mountain areas.

A number of geographers also attempted to investigate the depopulation trend, through intensive studies of the process of change in mountain villages, and they succeeded in identifying the relationships between out-migration and the changing social structure of the village community. Their results were not influential, however, in part because they confined their attention only to small areas, and despite all their efforts they were unable to offer a solution to the problems of these areas.

It is widely thought that depopulation was a direct consequence of the government's economic policies, which stressed large capital investments in heavy industries located in metropolitan areas. It must also be recognized, however, that the economic base of mountain areas was itself becoming much weaker over the same period. There were three stages, I believe, in this weakening process.

The first stage saw a collapse in the production of firewood and charcoal from 1960 onwards, due to the increased availability of cheap oil. This collapse destroyed the economic base of the southwestern mountain areas, which had depended heavily upon these products.

At the same time, however, mountain areas in the northeast were able to switch emphasis to the cultivation of paddy rice, thanks to the government's price support system. The second stage of economic weakening was caused by changes in this price support system from 1970 onwards which were designed to reduce rice production. Many farmers in the northeast were forced to resort to part-time jobs in metropolitan regions between autumn and spring. As a result depopulated areas were formed mainly on a seasonal basis, in winter.

The third stage developed after 1970, when the depopulation wave reached areas located near large markets, and which hitherto had traditionally specialized in forestry. Problems were caused here by the availability of large amounts of cheap imported timber, at a time when the price of Japanese timber was being driven up by the high rate of economic growth. As a result of this competition Japanese timber prices fell rapidly, and many people left their mountain villages to seek work in metropolitan regions instead.

As a result of these common factors which weakened their economic base, depopulated areas emerged all over Japan. And not surprisingly, given their economic collapse, these were also districts with per capita incomes well below the national average.

Regional policies for mountain areas as a short term solution, and resulting changes in mountain villages

In the early stage of the depopulation process, local governments were unable to stem the outflow of population to metropolitan areas. The mountainous parts of Chûgoku had been the centre of wood and charcoal production, and thus were affected dramatically by the shift to oil. In this area, the movement of population from mountain to city resembled a landslide. The national government had no policies capable of restricting this migration either. It recognized the problem too late, and besides, several leading economists had supported the rather optimistic view that migration from mountain areas was simply a sign of income readjustment between metropolitan and rural areas.

Yet the departure of former residents caused many problems in the communities affected: shortages of young labourers, the loss of community functions, an inability to promote conservation measures, more damage caused by animals, and so on.

The government introduced its first attempt at a policy, the *Sanson Shinkô Hô* (Mountain Village Development Law), in 1965. The purpose of this law was to supply funds to municipalities which met specified conditions, for use in depopulated areas under their juristiction. These conditions stated that more than 75% of the land area within the municipality had to consist of woodland, and that the average population density had to be below 116 people/km^2. A total of around 1,100 municipalities met these conditions. The law strictly limited the use of funds to a range of centrally-approved measures, which were limited in their effectiveness on account of their simplicity and through failure to tailor them to suit each area's specific problems. As a result, the law had little effect on mountain areas, and out-migration continued.

Many Diet members representing mountain areas became apprehensive over the likely results of national elections during this period, because of the rapid decrease in the number of voters in their constituences: they feared that this decrease would trigger a reduction in the number of Diet members elected from mountain areas. In response to these fears, new policy ideas were widely discussed by all the political parties, from the ruling Liberal Democratic party to the Communist party, with the outcome that the government had to introduce a new law to aid depopulated areas. This law, which had a ten-year life, came into force in 1970 under the name *Kaso Hô*.

This legislation was epoch making in as much as it permitted each municipality to issue bonds to fund public infrastructure, with the central government paying the interest on the bonds. The provisions of the law also reflected the insights gained by researchers who had questioned the factors underlying migration to urban areas, and who had emphasized the specific characteristics of remote and inconvenient locations. The law applied to any municipality which had suffered a population decrease of over 10% in 1960-65, and which provided less than 40% of its financial needs from local taxes.

The areas designated as depopulated by the government were distributed throughout Japan and covered about half the national territory, but their combined population was only 7.5% of Japan's total. Under the *Kasô Hô*, a great deal of money was fed into these areas to fund plans drawn up by the municipalities, and most was spent on infrastructure projects, such as main and forest roads, bridges, tunnels, public offices and halls, modern schools (to promote amalgamation), and so on. As a result, the landscape of mountain villages underwent noticeable changes.

These investments also created a substantial demand for labour in mountain areas, affording work opportunities to many of the elderly people in the villages. As a result, the number of local people engaged in public works increased to the point where this became the greatest source of income in many villages. Men and women who had formerly been engaged in forestry or agriculture switched occupations, though most became nothing more than day-labourers.

The contractors for these public works, however, were mainly drawn from urban areas. This was especially true for large projects, because local contractors in mountain areas were too small and their equipment inadequate. As a result much of the money fed into mountain areas soon returned to urban areas: that which remained consisted mostly of the wages earned by local workers.

At the same time, the government introduced a new regional policy to encourage manufacturing in rural and mountain areas through favourable tax treatment. Tax inducements persuaded a number of companies to build new factories in mountain areas, in industries such as ready-made clothing and electronic component assembly. These companies were small-scale, however, and only wanted cheap female labour. As a result, they did little to strengthen the economies of mountain villages. Besides, the 1973 oil shock triggered the closure of many of these manufacturing concerns.

After the oil shock, as the economic development of Japan slowed down, so the rate of population loss in mountain areas also slowed, but by then most of the young and middle aged people had already left. As factories closed down, so the opportunity to gain employment on local public works projects became even more important.

In response, the *Kasô* law was extended for another 10 years, to 1990, and was renamed the *Shin* (new) *Kasô Hô*.

Enforcement of the *Shin Kasô Hô* and the regionalization of mountain villages

The new law stressed that each mountain village should endeavour to develop its own economic base. This would be very difficult to achieve, but it was an important change of policy, for it meant a switch from trying to developing mountain villages simply by augmenting their infrastructure to strengthening them from within.

Since 1975, new marketing and circulation systems have developed in Japan, and local cities have been boosted by the addition of branch offices of enterprises headquartered in metropolitan areas. As a result, these cities have been able to attract more population, with consequent effects upon mountainous areas. Demographic data show clearly that municipalities in mountain areas which have achieved increases in population over the past decade tend to be located near prefectural capitals, especially in the "inner zone" of southwestern Japan. Access to these cities became easier thanks to new roads funded by the earlier *Kasô Hô* plan. This led to the construction of new factories operated by sub-contracting firms (mainly parts makers), and subsequently to the development of new residential areas in and around these cities.

On the other hand, municipalities which have continued to lose population are mostly located in the "outer zone" of southwestern Japan. It is difficult to build roads in these areas due to geographical conditions, and hence accessibility remains a problem. The population in many of these municipalities is heavily concentrated in the over 65 age group, and indeed, some villages consist only of old people. These villages will soon disappear entirely. In the past, when settlements vanished it was because of emigration to urban areas, but from now on we will also observe settlements that disappear simply because of ageing: soon there will be large "socially-formed empty areas" in the mountainous zones of Japan. These areas cannot survive unless people move in from urban zones, but there is little sign of such a trend.

"Socially-formed empty areas" are an important problem, in particular because they lead to the creation of waste land within mountain areas which are bound to face severe environmental problems. The government is expected to introduce a new regional policy for mountain areas in response, which is likely to include a new method of funding, in which expenditures will be allocated to these "socially formed empty areas" and not to the municipality as a whole.

Recently, the Japanese public has become more aware of the problems that mountain areas face, thanks to the activities of a forest preservation movement promoted by the major newspapers. At the same time, some mountain villages have tried to establish relationships with urban areas by selling them traditional local products. Village people hope to strengthen the economic base of their communities in this way through their own efforts, seeking their own solution to their problems. In addition, a fair number of people who once lived in urban zones are today moving to mountainous areas, in search of a new style of life.

As a result, although many of these mountain areas are facing disaster as a consequence of depopulation, there is still some hope for the future, for many are actively seeking solutions to their dilemma. As to how and when these new projects will be implemented, and how successful they will be, that remains to be seen.

Local Area Economic Fluctuations in Terms of New Job Offers In Northern Kyushu, Japan

Yoshitsuyu Asami
Kurume University

Introduction

The object of this study is to identify the spatial transmission patterns of economic impulses as manifested in new job offers, an indicator of local labour demand levels, within the Northern Kyushu district.

Research on regional economic impulses has been undertaken since the 1940's, mainly in the U.K. and the United States. Lösch (1962, first edition: 1940) and Vining (1946) were the first to conceptualize and analyse such impulses as a form of geographical diffusion. A group of researchers represented by King (e.g. King et.al., 1969) and Bassett and Haggett (1971) expanded this work to encompass the study of spatial diffusion and urban systems.

Since the early 1970s, several quantitative studies have explored the geographical patterns of economic fluctuations, the inter-area transmission of fluctuations and their associated lead-lag relationships, and the impact of national economic fluctuations upon specific areas (Asami, 1980; Kohsaka, 1984). The emphasis in this work has been on the quantitative analysis of economic impulses, but the findings have generally been interpreted in the context of the industrial composition of the area concerned. R.J. Bennett in particular analysed regional economic cycles using autoregressive moving average (ARMA) techniques. For example, in Bennett (1975) regional unemployment, regional population, industrial migration and other variables are modelled using a Kalman filter.

In the late 1970s and early 1980s, sophisticated techniques such as the Box-Jenkins method were introduced to examine the chain diffusion of economic fluctuations, while previous methods of analysis, models and the interpretations of findings were critically reexamined. Clark (1980) reviewed progress in this area, and paved the way for interpretations based upon labour market theory.

The following interpretations have been made of the spatial transmission of economic fluctuations. First, they have been analysed in terms of the industrial composition of areas or of the inter-area supply and demand of goods. Second, they have been examined in the context of the Phillips curve, the inverse relationship between unemployment and wage inflation. And third, as far as the relationship between unemployment and the structure of labour market is concerned, attention has been focussed on real wages, hours of work, and labour union organization in the areas concerned.

Thus, geographical research on area economic fluctuations originated at the end of the 1960s and initially concentrated on inter-area variations and relationships in fluctuation patterns. Subsequently, in the late 1970s, research laid more emphasis on interpretation and other considerations. In the process, however, the spatial perspective became rather neglected.

The Box-Jenkins Model

1) ARIMA model

The autoregressive moving average (ARMA) model is a stochastic time series model for a specific time domain. Box-Jenkins extended this model as an autoregressive integrated moving average (ARIMA) model, and developed the methods of calculation and associated computer programmes. In spite of the complexity of the method, the model has proved popular, especially in the field of econometrics, and also in some geographical studies (e.g. Clark, 1979).

The Box-Jenkins ARIMA model is expressed as ARIMA(p,d,q), where p is the order of the autoregressive (AR) process, q is the order of the moving average (MA) process, and d is the degree of difference required for the time series to be stationary. The ARIMA(p,d,q) process is expressed by the following equation:

$$\phi(B) \nabla^d X_t = \theta_0 + \theta(B) a_t \quad ... \quad (1)$$

where

$$\phi(B) = 1 - \phi_1 B - \phi_2 B^2 - \cdots - \phi_p B^p)$$
$$\theta(B) = 1 - \theta_1 B - \theta_2 B^2 - \cdots - \theta_q B^q)$$

and where X_t is the time series, $\phi_1 \cdots \phi_p$ are AR (autoregressive) parameters (p: order of AR), and $\theta_1 \cdots \theta_q$ are MA (moving average) parameters (q: order of MA). The autoregressive (AR) term expresses the level of dependence on the accumulated previous values of X_t, and the moving average (MA) term expresses the level of dependence on accumulated present and past random noise a_t. B is known as a backward operator, and represents the backward shift of time period t, with $B^k X_t$

being equivalent to X_{t-k}. ∇^d represents the d-degree difference, and a_t represents the existence of random inputs at present and past levels. Finally, θ_0 is the overall constant, representing 'stochastic trends' as random changes in the level or slope (Box and Jenkins, 1976, pp. 9-12, 92).

2) A seasonal model

Most economic indicators have four components: a linear trend corresponding to economic growth, a cyclical component linked to business cycles at intervals of 3 or 10 years, a seasonal component determined by the seasonal cycle, and an irregular component specific to each series.

Box and Jenkins (1976) have presented a seasonal model for a time series displaying seasonal cycles. The seasonal ARIMA model, which duplicates the seasonal terms, is expressed as ARIMA(p,d,q)(P,D,Q), where P is the order of the seasonal-AR process, Q is the order of the seasonal-MA process and D is the degree of seasonal difference. The seasonal ARIMA(p,d,q)(P,D,Q) process is expressed by the following equation:

$$\phi\,(B)\,\Phi\,(B^s)\,\nabla^d\,\nabla_s^D\,X_t \;=\; \theta_0 + \theta\,(B)\,\Theta\,(B^s)\,a_t \quad \dots \quad (2)$$

where

$$\nabla_s^D = (1 - B^s)^D$$
$$\Phi\,(B^s) = 1 - \Phi_1\,B^s - \Phi_2\,B^{2s} - \dots - \Phi_P\,B^{Ps}$$
$$\Theta\,(B^s) = 1 - \Theta_1\,B^s - \Theta_2\,B^{2s} - \dots - \Theta_Q\,B^{Qs}$$

and where s is the period of seasonal cycle, ∇_s^D is the difference for X_t and X_{t-s}, $\Phi_1 \dots \Phi_P$ are the parameters of the seasonal-AR terms (P: order) and $\Theta_1 \dots \Theta_Q$ are the parameters of the seasonal-MA terms (Q: order). The backward operator B^s represents the backward shift between time periods t and t-s. For example, $B^k\,{}^sX_t$ is equivalent to X_{t-ks}. ∇_s^D represents the D-degree difference at interval s.

3) Modelling procedure

i) Model identification

The order of the ARIMA process (p,d,q) is determined from the patterns of the autocorrelation-function (ACRF) and the partial autocorrelation function (PACF), by referring to the theoretical ACRF or PACF functions to determine p, d and q.

ii) Estimation of parameters

The parameters of the model are estimated by non-linear estimation methods. Box and Jenkins (1976) used the Marquardt algorithm, through which the residual sum of squares is minimized. A preliminary estimate is made to speed up the process of calculation.

iii) Diagnostic checking

The estimated models are checked to determine whether the latter is adequate or not. If the residual time series after estimation does not have autocorrelations, then the residual series is assumed to be random white noise, and the model is adopted. If this series has autocorrelations, however, the model is deemed to be inadequate and is rejected. If the model is rejected, the order of the model, p, d, q, P, D and Q must be changed and the model re-estimated. Chi-square tests and Box-Pierce statistics are used for checking.

Data and study area

1) New job offers
In this paper new job offers have been used as an indicator of regional economic fluctuations. They are reported by each Public Employment Security Office (PESO) on the basis of vacancies notified by establishments within its jurisdictional area. This indicator is meaningful in as much as new job offers (or labour demand) are part of the process of employment increase, which is itself a reflection of the level of economic activity. In addition, this indicator is available at a local area level, consisting in most cases of a medium or small scale city and its surrounding rural area (approximating to a daily urban system). The metropolitan area is further divided into sub-areas. Few other time series are available at such a high level of spatial resolution in Japan.

Most of the new job offer series include a distinct seasonal component, with two bottoms in June and December. Figure 1 shows the new job offer series for the period from January 1968 to December 1985 for Fukuoka prefecture, which covers most of the study area. The irregular line indicates the raw series, while the smooth line shows the seasonally adjusted series. Seasonal adjustment has been achieved through the use of the moving average method, which is rather different from the moving average process mentioned above. The moving average was calculated with two steps, at 12 months and 2 months, as shown by the following equation:

$$X'_t = (X_{t-6} + 2 X_{t-5} + 2X_{t-4} + \ldots + 2X_t + \ldots + 2X_{t+5} + X_{t+6}) / 24 \ldots (3)$$

The pattern of fluctuations in this series is similar to the typical pattern of movement in new job offers for Japan as a whole. Cyclical fluctuations can be observed within the seasonal adjusted series, with periods of 3 to 6 years. These fluctuations correspond to the reference dates of the business cycle turning points determined by Japan's Economic Planning Agency. The highest peak corresponds to the upsurge of labour demand at the end of Japan's long postwar era of rapid economic growth. The recession that followed was caused by the first oil shock. From that time on, Japanese economic performance switched from high growth to stable growth.

In this study, new job offer series were analysed for the 11 year (132 month) period from January 1975 to December 1985. This period covers the era of stable economic growth after 1974, during which temporary employment subsidies have been in effect in former coal mining areas. Throughout these 11 years, the performance of this series has been relatively stable, close in fact to the stationary state.

2) Study area

The study area, Fukuoka prefecture and some adjoining municipalities (Figure 2), is located in Northern Kyushu. From the late 19th century to the 1970s this area was one of the most prosperous regions in Japan, specializing in coal mining and heavy industries, especially iron and steel. By 1973, however, coal mining had been abandoned completely save for one area, the Miike coalfield. Heavy industries became less prosperous after the first oil shock of 1973, and this became the most economically depressed area in Japan. However Fukuoka City, the central city of the Kyushu district, has enjoyed increasing prosperity as a provincial metropolis. The study area consists of 17 sub-areas, each the juristictional area of a particular PESO, and the data employed are the raw and seasonal adjusted series for each sub-area.

Patterns of fluctuation by area in terms of the ARIMA model

1) Raw series and seasonal adjusted series

In analysing the new job offer series we have adopted the seasonal model mentioned above, because these series have strong seasonal components. Asami (1988) has tried to fit non-seasonal ARIMA models to these series, but with limited success. For the seasonal model estimated here, the following results were obtained.

Three types of fluctuations were found amongst the raw and seasonal adjusted series for the various subareas. The first type includes series which have troughs in December (and June). Fukuoka, Omuta, Yawata, Kurume, Kokura, Kashii and Amagi belong to this type. These are mainly commercial and industrial cities, with relatively large scale labour markets. Their seasonally adjusted series have cyclical patterns corresponding to national cycles with some lead or lag. The second type is characteristic of Chikuho, the former coalfield area, and includes Iizuka, Nogata and Tagawa. Their series have one or two peaks a year, in April and October. These peaks can be attributed to the temporary employment measures in effect in this area. Cyclical fluctuations are relatively weak. The third type displays irregular fluctuations. Five members of this type, Tobata, Moji, Yukuhashi, Tosu and Arao have obvious cyclical fluctuations corresponding to national cycles but the others, Wakamatsu and Yame, have irregular cycles. These areas are characterized mainly by transport and manufacturing. Figure 3 shows the raw series and seasonally adjusted series for a representative city from each of these three groups.

2) Identification

These series are not always stationary, either in their non-seasonal or seasonal form. Thus, all series were differenced, the seasonal and non-seasonal series in turn. The ACRF and PACF were calculated for the raw series (X_t), the 1-degree differenced series (∇X_t), the 1-degree seasonal differenced series ($\nabla_{12} X_t$) and the 1-degree seasonal and non-seasonal differenced series ($\nabla \nabla_{12} X_t$).

These ACRF and PACF patterns were compared to the theoretical patterns for some ARIMA processes. We have selected some processes which are likely to resemble the theoretical pattern, estimated the parameters and undertaken diagnostic checks.

3) Estimation

Parameter estimation and diagnostic checking were undertaken using the methods described above. Table 1 shows the estimated models for the cities which have been chosen to represent the three groups, and Table 2 shows the identified processes for all of the series that were found to be significant after diagnostic checks. Significant models could be estimated for most of the series, except those for the Omuta sub-area.

Despite the fact that there was not much pattern in the raw and seasonal adjusted series, too many forms of the model were estimated. In particular, in the case of several sub-areas, more than one model yielded significant estimates from one data

58

series. In the main, three types of model were estimated. One is the non-seasonal AR(AR) and seasonal AR(SAR) type, which is characteristic of Fukuoka, Kurume, Kokura and Kashii, places which also belong to the first of the fluctuation type groups described earlier. The second type of model is the non-seasonal MA(MA) and SAR, as found for Iizuka, Nogata and Tagawa, places which belong to the second fluctuations type group. These places have relatively random raw series and smooth cyclical seasonally adjusted series. The third type of model is the AR and seasonal MA(SMA), characteristic of Tobata, Yukuhashi, Tosu and Arao. They have irregular seasonally adjusted series. In addition, several minor patterns were found.

Concluding remarks

This study aimed at analysing spatial structures in terms of the pattern of economic fluctuations. The Box-Jenkins method was adopted, a technique for estimating stochastic time series processes. Various models could be estimated significantly, and in some cases several models could be estimated from a single time series. The pattern of parameters by sub-area was very complicated, and the explanation of the meaning of these parameters is very difficult. Nevertheless, this method appears to be an effective approach to modelling the spatial process associated with economic activities.

References

Asami, Y., 1980, 'Leads and lags of economic fluctuations in terms of new job offers: The case of the northwestern part of the Tokyo metropolitan area', *Geographical Review of Japan*, 53, 329-344 (in Japanese).

Asami, Y.,1988, 'The analysis of regional fluctuations in terms of the new job offers in Fukuoka area: using Box-Jenkins ARIMA models', *Computer Information, Kurume University*, 4, 46-64.

Bassett, K, and Haggett, P., 1971, 'Towards short term forecasting for cyclic behaviour in a regional system of cities', in Chisholm, M, Frey, A.E., and Haggett, P. (eds.), *Regional Forecasting*, London: Butterworth, 389-413.

Bennett, R.J., 1975, 'Dynamic systems modelling of spatial systems', *Environment and Planning A*, 7, 525-538, 539-566, 617-636, 887-898.

Box, G.E.P., and Jenkins, G.M., 1976, *Time Series Analysis: Forecasting and Control*. (2nd. Edition). San Francisco: Holden-Day.

Clark, G.L.,1979, 'Predicting the regional impact of a full employment policy in Canada: A Box-Jenkins approach', *Economic Geography*, 55, 213-226.

Clark, G.L., 1980, 'Critical problems of geographical unemployment models', *Progress in Human Geography*, 4, 157-180.

King, L.J., Casetti, E., and Jeffrey, D., 1969, 'Economic impulses in a regional system of cities: A study of spatial interaction', *Regional Studies*, 3, 213-218.

Kohsaka, H., 1984, *Regional Economic Analysis: Spatial Efficiency and Equality*. Tokyo: Kobundo (in Japanese).

Lösch, A., 1962, *Die räumliche Ordnung der Wirtschaft (3 Auflage)*, Stuttgart: Gustav Fischer Verlag.

Vining, R., 1946, 'The region as a concept in business cycle analysis', *Econometrica*, 14, 201-218.

Table 1: Identified Process and Estimated Models for Typical Series (only models shown to be significant through diagnostic checking are displayed)

Area	Identified Process	Estimated Model	Residual Variance	BP[2]	DF[3]	ACR Reject
Kokura	(2,0,0)(1,1,0)	$(1 - 0.29B - 0.40B^2)(1 + 0.48B^{12})(\nabla^{12}X_t + 9.58) = a_t$ $(\pm0.09)\ (\pm0.08)\qquad (\pm0.08)$	30386.	68.31	56	0.13
Tagawa	(0,0,0)(1,0,0)	$(1 - 0.94B^{12})(X_t - 41.37) = a_t$ (±0.01)	46786.	55.24	58	0.58
	(0,1,1)(1,0,0)	$(1 - 0.91B^{12})(\nabla X_t + 0.04) = (1 - 0.99B)a_t$ $(\pm0.01)\qquad\qquad\qquad (\pm0.01)$	47520.	52.28	57	0.65
Tosu	(3,0,0)(0,0,0)	$(1 - 0.32B - 0.02B^2 - 0.30B^3)(X_t - 63.93) = a_t$ $(\pm0.08)(\pm0.09)\ (\pm0.08)$	2301.	72.56	56	0.07
	(0,1,1)(0,1,1)	$(\nabla\nabla^{12}X_t + 0.02) = (1 - 0.68B)(1 - 0.87B^{12})a_t$ $(\pm0.07)\quad (\pm0.03)$	2286.	44.80	57	0.88

1) **Identified Process**: ARIMA(p,d,q)(P,D,Q), where p,q and d are non-seasonal AR and MA parameters and the degree of difference respectively, and P, Q and D are the seasonal equivalents.
2) **BP**: Box-Pierce statistic.
3) **DF**: Degrees of freedom
4) **ACR Reject**: the significance level of the residual autocorrelation (ACR) function. If this level exceeds 0.05 the ACR is judged not to be significant and is rejected. If the ACR is rejected, then the estimated model is judged to be correct and is adopted. Only models which have been adopted are shown in this table.

Table 2: ARIMA Process - Seasonal Model*
(only models shown to be significant through diagnostic checking are displayed)

* ARIMA(p,d,q)(P,D,Q), where p,q and d are non-seasonal AR and MA parameters and the degree of difference respectively, and P, Q and D are the seasonal equivalents. ∇ within parentheses means difference. Superscripts refer to the order of difference, subscripts to the period of difference. ∇^1 means a non-seasonal difference, and ∇_{12}^1 a seasonal difference.

Area	$(\nabla^0\nabla_{12}^0)$	$(\nabla^1\nabla_{12}^0)$	$(\nabla^0\nabla_{12}^1)$	$(\nabla^1\nabla_{12}^1)$
Fukuoka	(1,1,1)(1,1,0)			
Kurume			(3,0,0)(1,1,0)	
Kokura			(2,0,0)(1,1,0)	
Kashii			(2,0,0)(1,1,0)	
	(0,1,1)(0,1,1)			
Yame			(2,0,0)(1,1,0)	
	(0,1,1)(1,1,0)			
Tagawa	(0,0,0)(1,0,0)	(0,1,1)(1,0,0)		
Nogata	(0,0,0)(1,0,0)		(0,0,0)(1,1,1)	
	(0,1,1)(1,1,0)			
Iizuka	(0,0,0)(1,0,0)		(0,0,1)(0,1,1)	
Arao	(2,0,0)(0,0,0)	(0,1,1)(2,0,0)	(2,0,2)(1,1,1)	
Tobata			(2,0,0)(1,1,1)	
Tosu	(3,0,0)(0,0,0)			
	(0,1,1)(0,1,1)			
Yukuhashi	(1,0,0)(0,0,0)	(0,1,1)(0,0,0)	(1,0,0)(1,1,1)	
	(0,1,1)(0,1,1)			
Wakamatsu	(1,0,1)(1,0,1)	(0,1,1)(0,0,1)		
	(0,1,1)(1,1,0)			
Moji	(0,0,1)(1,0,0)	(0,1,1)(1,0,0)	(1,0,0)(1,1,1)	
	(0,1,1)(1,1,0)			
Amagi	(0,0,0)(1,0,1)	(0,1,1)(1,0,0)	(0,0,1)(1,1,1)	
Yahata				
	(0,1,1)(0,1,1)			
Omuta	- no significant models -			

Figure 1: New Job Offers in Fukuoka Prefecture, 1968-1985

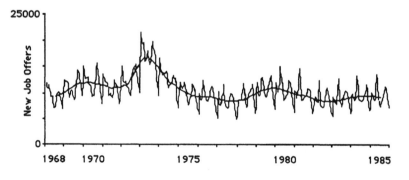

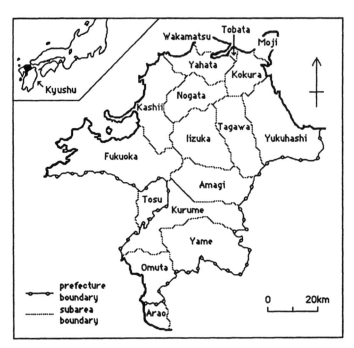

Figure 2: Study Area

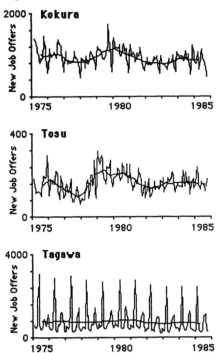

Figure 3: Some Typical Examples of Raw and Seasonally Adjusted Series

Part III: Urban Geography

Pollution and Neighbourhood Organizations in Ojima, Kôtô-ku, Tokyo: A Progress Report

Dusan Simko
University of Basle

Since the OECD published its first review of environmental policies in Japan in 1977, the image of Japan with regard to pollution has changed. In the early seventies Japan gained notoriety as a country willing to commit "ecological suicide" for the sake of economic growth, yet in the later seventies it came to be considered a model for Western European countries in terms of effective pollution control. The reasons for this reassessment of Japan's situation include improvements in some aspects of air pollution, especially the reduction of sulphur dioxide concentration in the air.

Prompted by the prospect of forests dying from acid rain and an increasing incidence of chronic bronchial disease amongst children in major Swiss cities (e.g. Basle, Zurich and St. Gallen), both of which are due to inadequate environmental policies, part of my research has addressed the question of whether the Japanese experience can really serve as a model in the field of pollution control. Hence I have concentrated partly on how Japanese political and administrative bodies have succeeded in obtaining the cooperation of firms in technological problem-solving through pollution abatement facilities. It is at the local level that environmental problems first become apparent and complaints and demands addressed to politicians arise most frequently. It is also at the local level that the chances of flexible problem-oriented solutions seem to be greatest, because of the intimate knowledge there of the local problem structure. My study therefore deals with local government policies in Tokyo and with the social response of neighbourhoods.

It is well known that many of the older parts of Japanese cities are divided into well-defined neighbourhoods. When urban neighbourhoods are discussed, the debate usually centres on the significance of the *chônaikai*, or neighbourhood associations, which are key institutions in the formal structure of many neighbourhoods. Because of the important role played by *chônaikai* and by related organizations in local government and politics, they have received close attention from scholars interested in recent Japanese history and politics. These scholars usually view the social relations and activities of the neighbourhoods in political and administrative terms. But the *chônaikai* are much more than an extension of the municipal or ward government; they provide social control, information, social solidarity, and ensure consistency. Indeed, some authors consider the *chônaikai* to be descended from the old rural neighbourhood institutions of preindustrial Japan.

The most important of the formal neighbourhood organizations are the semi-voluntary "common interest associations" such as the *chônaikai* and their womens' auxiliary associations (*fujinbu*), the senior citizens' club (*rôjinkai*), the festival committees (*saireibu*), the merchants' associations (*shôtenkai*), PTA (Parents Teachers Association), politicians' support clubs (*kôenkai*), and the volunteer fire brigades (*shôbôdan*). The most "powerful" and obvious of them all are the *chônaikai*. The *chônaikai* ensure that a flow of information is established between the authorities and the neighbourhood, and the different types of *chônaikai* committees provide a variety of services to the residents. Information to residents on government programmes and regulations, help with censuses, and conducting surveys of local conditions are among the activities undertaken by the *chônaikai*, and the local branch of the ward office considers the *chônaikai* to be under its jurisdiction. The Ward office refers to the *chônaikai* collectively as a *"burokku"* (block) under its leadership, while the *chônaikai* leaders see themselves as members of a federation supported by the local ward office. Thus it often happens that the local ward office plans and sponsors a number of traditional activities that duplicate the efforts of the *chônaikai*.

Ojima: A Case Study

The Ojima neighbourhood in northeast Kôtô-ku (Kôtô Ward) was established during the prewar industrial development of Tokyo, and thus did not originate as either a rural hamlet or a traditional merchant quarter.

Social and cultural patterns in Kôtô-ku have become homogeneous over the past few decades because of massive social and economic mobility, the rapid influx of migrants into major urban centres, and the encroachment by the growing central business district, factors which have helped to change the structure of Tokyo's *shitamachi*. Thus, while Nihombashi is now largely a central business district, areas such as Asakusa and Ueno retain their *shitamachi* aura, while newer areas such as Kôtô, Arakawa and Sumida wards have acquired many *shitamachi* characteristics. Here in these "non-classical" *shitamachi* areas of Tokyo, neighbourhood social life preserves much of the open informality that contrasts so noticeably with the more reserved people of *yamanote*.

Kôtô ward is located between the Sumidagawa and Arakawa rivers. After the catastrophes of 1923 and 1945, Kôtô-ku recovered each time to become a busy commercial and industrial district. The principal industries are metals, machinery, chemicals, foodstuffs and lumber. The ward had a population of 387,479 persons in 1987, an increase of around 7% over the prevous five years, living in an area of 36.89 sq km. In Ojima itself there were some 59,821 persons living in 22,188 households. General statistics on Kôtô-ku indicate that the largest number of inhabitants are immigrants from other parts of Tokyo. The neighbouring prefectures of the Kantô Plain, Chiba, Tochigi, Saitama and Ibaraki, are the next most important source of immigrants, followed by the northeastern part of the main island of Honshû, Tôhoku.

★★★★★★★★★★★★★★★

The 200 families sampled in Ojima more or less fit this general pattern. Only about a quarter were born in other parts of Tokyo, mostly in places within the old *shitamachi*, such as Asakusa, Shitaya and Kyôbashi. The distribution of occupations in our sample reflects the lower and lower middle class composition of Kôtô-ku generally. Of course, it would be misleading to describe this area as being of generally low class, for its heterogeneous social composition is very different from the well-known social homogeneity characteristic of Central European cities. Many owners of medium-size factory units in Kôtô-ku still live in compounds adjacent to the production site. Although the big factories pulled out of the Ojima district in the seventies, the medium- and small-size companies have resisted such a move. Many of the small firms in Ojima are family-based and act as subcontractors: here the typical combined home and factory, with fewer than ten employees, still survives. This structure has tended to hinder changes in land ownership within our survey area.

We also personally interviewed 60 shopkeepers who were members of the Ojima-based shop owners' association (*shôtenkai*). We also conducted long interviews with the head of the local *chônaikai* (*kaichô*), Kumihiko-san, the head of the local *shôtenkai*, and the *Ojima jinja gunji* (Master of the Higashi Ojima Shrine), Tomihiko-san.

After 200 questionnaires had been returned, we visited 30 families in their homes to speak to the wives or - less frequently - to the husbands about their attitudes towards the environmental problems of their neighbourhood.

The *chônaikai* were revived in Kôtô-ku - and in Ojima - in 1951, six years after the end of World War II, when the earlier *tonari gumi* and *chôkai* had been dissolved. Membership of the *chônaikai* is voluntary in principle, but in the seventies social pressure in Kôtô-ku was probably sufficient to bring nearly 95% of all households into the *chônaikai*. Today in Ojima 1,360 households (equivalent to around 4,300 persons) are members of the local *chônaikai*. This indicates a certain decline in the social importance of the *chônaikai* in our survey area.

Of the Ojima *chônaikai* members, 280 are over 60 years old. Every household member of the *chônaikai* pays between 15 and 100 yen monthly. The members elect the president (*kaichô*), who in turn appoints two vice-presidents (*fukukaichô*), two treasurers (*kaikei*) and two supervisors (*kansa*). The *chônaikai* have different sections - such as a traffic section (*kôtsubu*), a general affairs section (*sômubu*), a women's section (*fujinbu*), a crime prevention section (*bôhanbu*), a welfare section (*kôseibu*) and the youth section (*seinenbu*). Each section is headed by an elected director. The leading officials of the *chônaikai* hold regular meetings of all members once a month - in the private home of the *kaichô* - to discuss and decide upon activities. A general meeting of all members is held once a year in the ward hall auditorium. The *chonaikai* office is located in a large house (11 rooms) owned by the *kaichô* of Ojima, a former factory owner (age 75).

Thanks to the *chônaikai* a very large network of individuals is given tasks and titles related to the activities at hand: For example, the *chônaikai* takes the initiative in maintaining a clean and neat neighbourhood, by encouraging people to sweep the streets, supervising local refuse collection by the ward, and campaigning for reductions in the use of detergents. The *chônaikai* may also submit requests to the ward office and municipal government for the repair and improvement of streets and bridges, the installation of traffic lights and road signs etc., or for public safety measures. In cooperation with the police and fire stations, the *chônaikai* assumes responsibility for maintaining order and safety in its neighbourhood - for example in, organizing the *keibôdan* (a vigilante corps of civil guards), whose members take turns going the rounds on winter nights beating wooden clappers (*hi no yojin*). Another function of the *keibôdan* is to help children to cross streets in heavy traffic near the subway stations in Ojima.

The *shôtenkai*, or shop owners' associations, comprise all the owners of shops, restaurants and other retail businesses in the same neighbourhood, regardless of the nature of their services. The criterion of local is absolute. For instance, a shopkeeper who has his business in Ojima and resides in Kameido will belong to the shôtenkai of Ojima and the chônaikai of Kameido. Internal organization is much the same as in the chônaikai. The main function of the shôtenkai is to facilitate its members' business by cooperative action, such as holding special sale days, decorating the shops for special seasonal events such as the New Year, and issuing raffle tickets to their customers.

The urban environment of Ojima lies mainly below sea level and is partly overcrowded with low rise wooden houses. In view of its vulnerability to disasters, drastic disaster-prevention measures for the entire district are required. In response, the Tokyo Metropolitan Government approved the Kôtô Redevelopment Basic Concept, under which it decided in November 1969 to construct six large-scale fortified shelter bases as a part of its comprehensive redevelopment plan for the area. Development of the shelter bases in the Shirahige, Kameido, Ojima and Komatsugawa districts has been undertaken as a built-up area redevelopment project. The city plan for Kameido, Ojima and Komatsugawa was drawn up in 1975, and construction work was started in 1980 and completed in 1988. With this development the government hopes to improve the living environment for the entire district. In the event of a major earthquake the residents should have immediate access to evacuation sites. The new comprehensive plan for the Kameido, Ojima and Komatsugawa district calls for comprehensive water control measures, i.e. flood control and protection from flooding caused by tidal waves, within the so-called zero-meter areas of Kôtô-ku and the eastern part of Tokyo. Improvement of the drainage system in the ward area has made steady progress in recent years, and satisfied the needs of 67% of the population of Kôtô-ku by 1985.

As can be seen from the air pollution reports, environmental quality standards for carbon monoxide and sulphur dioxide have already been established, and one for nitrogen dioxide is in preparation, but standards for photochemical oxidants and suspended particulate matter have not yet been defined, and effective measures are needed. The situation with respect to the rivers and sea water has shown a general improvement, but there has been a slowdown in this trend in recent years. As a result, the further purification of river water and the prevention of eutrophication in Tokyo Bay have become big problems. The cooperation of citizens is particularly essential with regard to phosphorus reduction. In addition noise pollution, especially traffic noise along the arterial trunk roads, local noise from small-scale factory units and shops, offensive odours, soil contamination etc. are all problems which need to be solved in the future.

In order to create a safe and comfortable environment, it is necessary to establish and implement a realistic Environmental Control Plan which not only controls the various causes of pollution in cooperation with local industries, but also takes into account local characteristics, and which improves the information control system between the community level, the Environment Agency (Kankyô-cho) and its own research organization, the National Institute for Environmental Studies, and the major projects overseen by the Ministry of International Trade and Industry. (It is a paradox that this ministry is responsible for environmental research). In the 1970's Japan adopted the "Polluter Pays" Principle according to which enterprises causing pollution had to accept financial responsibility for the damage they inflicted on the community. Even so, the tolerance limits for many substances remained high, and it became clear that when environmental goals conflicted with "stable" growth, the latter would prevail.

Finally, I would like briefly to summarize the questions that formed the main focus of our investigations in the Ojima district:

1. Do social characteristics account for differences in attitudes towards pollution?
2. Do attitudes toward pollution depend on the level of exposure to pollution?
3. Can differences in attitudes towards pollution abatement be explained in terms of social characteristics?
4. Are attitudes towards pollution abatement a function of the pollution level in the area?

Evaluation of the interview material is still in progress, but we may already draw the following conclusions:

a. Community attitudes can be differentiated in terms of social characteristics, as follows:

a.1. Persons in white-collar occupations are less concerned about and feel less bothered by pollution than blue-collar workers.

a.2. *Danchi*-dwellers are more bothered by pollution and find it a more serious problem than homeowners do.

a.3. Persons in high-value homes show less concern about pollution than do those who live in low-value homes.

a.4. Community willingness to control pollution is not a function of the pollution levels for that area.

These are just a few of the many points raised by our investigations. We hope that answers to these questions will be forthcoming shortly, as well as opportunities to apply this type of analysis to the formulation of policy.

It will also be necessary to establish ways for Japan's citizens to participate in the process of pollution control, for administrative guidance and agreements made "behind closed doors" between bureaucracy and industry, and in favour of the latter, will only serve to increase the potential for conflict at the local community level.

Territoriality and Public Participation in the Process of Budget Planning in Narashino City

Gen Ueda
Hitotsubashi University

The research described in this paper is designed to illuminate the characteristics and problems associated with the system of public participation in local government in Japan. This system is a combination of two elements: the "area approach" to urban management adopted by the municipality, and the participation of the area residents themselves in the decision-making process. Japan tends to have a low level of citizens' involvement, and therefore its system of participation merits careful attention.

The specific focus is on the system of public participation in Narashino City, Chiba Prefecture. The system in Narashino was established under a reformist local government and has survived for almost twenty years. Yet the most serious problem seems to be that even now few citizens actually know of the existence of the system itself. To understand why this is the case, it is necessary to investigate the process through which participation has become systematised, irrespective of the potentially different motives which citizens may have had for participation once the institutions were in place.

In Narashino it was local government rather than the citizens which took the first initiative, by organizing a territorial system for participation based on the "area approach", in which public officials were allocated to each "community" as defined by the local government itself. The local government also stimulated the formation of a voluntary citizens organization linked to its own territorial system, in an attempt to facilitate citizen-centered control of the life space. This attempt was opposed by existing neighbourhood organizations (*chôkai* and *jichikai*), however, because the administrative boundaries of the districts used as the basis for the new voluntary citizens organization did not coincide with those of the existing organizations.

The documentation of a conference of citizen representatives on the process of territorialization, which was held on the initiative of the local government, has been analyzed using a type of canonical correlation analysis. The analysis identifies three different groups of opinions regarding the best procedure for promoting the management of life space: Group I, which called for participation on the basis of existing neighbourhood organizations, to which people living within those areas belong almost automatically anyway; Group II, which wanted to organize non-territorially motivated citizens as well, and who were not concerned to act solely for the benefit of a specific neighbourhood organization; and Group III, which had no intention of participating in municipal decision-making, but wished instead to promote a true "grass roots" movement. Group I was the most dominant at the conference in terms of numbers. Moreover, it enjoyed the support of local assemblymen, who criticized the attempts of local government to systematize participation independently of the neighbourhood organizations. As a result, the problem of systematizing participation was eventually settled by linking the existing territorial organizations of citizens to the new territorial system set up by the local government. As a result, existing neighbourhood leaders retained a kind of *de facto* territorial right to represent the people organized under them within the new system.

This pattern of territorialization had several effects: it obscured the definition of representation by neglecting to differentiate between the right of neighbourhood organizations to represent their members and the rights of non-territorially motivated volunteers; it excluded non-organized citizens from the system of participation; and it served to channel citizen demands on the local government arising at levels below that of the "community". There was a territorial ideology present in the arguments put forward by the citizens, an ideology which rationalized the obscuring of the definition of representation, for the citizens did not recognize any necessity for non-territorially based volunteers to participate, postulating instead that "all" the people within the area of a neighbourhood organization should participate through that organization. This ideology seems to have been shared by some public officials as well. Territorialization in Narashino, therefore, was a politico-spatial process of conflict among volunteers, the leaders of neighbourhood organizations and the local government over how life space should be managed.

An attempt has been made to evaluate the effectiveness of this system in supporting broad participation within the city's sixty-seven districts, for a period of six years starting from 1977, the year in which the system finally became fully established in its present form. For this purpose three variables have been used. The first variable is the rate of citizen enrolment in neighbourhood organizations, and has been chosen on the assumption that a high rate of enrolment is a necessary condition for broad participation within the system, given the importance of neighbourhood organizations in the territorialization process. The second variable is the total number of demands made by citizens on local government each year through the system of participation, which reflects the reality of actual participation in the budget planning process. The coefficient of variation for the number of demands over this period is the third variable, and has been chosen as a measure of the stability of demands. In evaluating the effectiveness of the system of participation, it is meaningless to emphasize the actual budget allocations agreed relative to the number of demands made, partly because some demands are dealt with through negotiations and the efforts of government officials without any payments being made, and partly because the sums budgeted do not necessarily reflect the benefits subsequently enjoyed by the citizens affected. This is the reason why the absolute number of demands has been selected as a variable, in preference to sums budgeted.

These variables have been used to cluster the sixty-seven districts into eleven groups. The effectiveness of the system within each group has then been evaluated, a process which has resulted in the further classification of the eleven groups into four types. There are three "effective" groups of districts, which have achieved a high percentage enrolment in neighbourhood organizations and generate numerous demands on local government, two "biased" groups, which communicate many demands but have a low degree of organization, two "ineffective" groups, which are also poorly organized and convey few demands, and four "latent" groups, which retain a high level of organization but issue a relatively small number of demands: these have the potential to become "effective" in the future. This breakdown reveals substantial differences in the degree of effectiveness between groups, differences which can be interpreted in terms of a structure of restricted participation, in which the "biased" and "ineffective" groups shows a low degree of organization associated with rapid urbanization, while the "effective" and "biased" groups are able to present a consistently high number of demands. This analysis has not been designed to reveal any spatial patterns in the effectiveness of the system of participation however.

In the next stage of the investigation, the same classification of districts was used in a canonical discriminant analysis in conjunction with a range of socio-economic variables derived from the 1980 census. The variables included population, age, household, residence, occupation, education and housing conditions. The analysis reveals three dimensions (canonical variates) which shed light on some of the differences between the eleven groups already defined in terms of their effectiveness in supporting broad participation. The first canonical variate picks up differences in the duration, size and ownership of residences, the second is distinguished by a relatively high percentage of older residents and the occupational dichotomy between secondary and tertiary industries, and the third is an index of the type and stability of households. This three-dimensional coordinate system shows that the groups which have been identified by other means share similar socio-economic characteristics. All of the "effective" groups and the "latent" group which comes closest to qualifying as "effective" share a high degree of owner-occupancy and household stability. In contrast, the "biased" and "ineffective" groups suffer from high rates of residential turnover. In addition, the "ineffective" groups have higher percentages of older residents and of persons engaging in tertiary industry. The other "latent" groups do not share similar characteristics however; they are scattered at various points in the coordinate system.

An appropriate question to ask at the level of the "community" is the extent to which the content of demands upon local government reflects the residents' evaluations of environmental conditions and public facilities. To answer this question, the multiple-group method of factor analysis was applied to a set of data on environmental evaluations derived from an inquiry conducted by the local government in 1984. The sixteen oblique principal components obtained were then correlated with the contents of the citizens' demands. The analysis reveals that the contents of demands communicated through the system of participation largely correspond to people's evaluation of life space. There is a clear distinction between the high valuations typical of areas located on reclaimed land and the low ones found in older areas. Older areas, which contain many "biased" and "ineffective" districts, also tend to submit specialized demands for improvement in public works in their immediate vicinity, whereas in newer areas demands relating to the whole area of the "community" are more common.

Thus, the system of public participation in Narashino has become just another manifestation of citizen's localism, in spite of the intentions of local government, and the degree of citizen involvement in the system depends very much on the actions of neighbourhood organizations. Moreover there is no mechanism to enable citizens to handle problems from a "community"-based or even broader viewpoint, or to make city-wide decisions. This problem is especially important for the older areas, which contain a number of "biased" and "ineffective" districts which submit many localized demands. In order to make the system of public participation meaningful for every citizen, it is necessary to correct these defects in the system, and to clarify the definition of citizen representation within it.

Part IV: Theoretical and Physical Geography

The Fractal Dimensions of Spatial Structures in Central Place Theory

Masayoshi Kikuchi

Toyo University

There is a close relationship between the hierarchical structure of central place systems with geometrical self-similarity and fractal theory. In this paper an extended fractal dimension is defined and the fractal dimensions of various central place systems are ascertained. The results differ from those obtained by Arlinghaus (1985) using the similarity dimension. An alternative procedure for generating a series of central place systems is proposed which preserves self-similarity in systems of the Christaller type.

1. Introduction

The theory of fractal geometry pioneered by Mandelbrot (1986) has penetrated various fields to describe irregular and complicated shapes, such as coastlines, mountains and rivers. The remarkable features of such shapes are that they are nowhere differentiable (ie. they are nowhere smooth) and that they possess the property of self-similarity.

Arlinghaus (1985) has suggested an application of the theory of fractals to the hierarchical structures of central place systems - systems which share this geometrical property of self-similarity. She proposes a method for constructing Christallerian central place systems of any K-value, where the K-value is the so-called Löschian number derived from the K-value generating function (Dacey, 1965), and for obtaining the fractal dimension of each central place system. Her method appears to create new difficulties, however, particularly in the procedure for generating the hierarchical structures of K-value central place systems through fractal iteration sequences and in the derivation of the associated fractal dimensions.

Some of these problems are discussed in Section 2 of this paper in the specific context of the Christaller model. An extended fractal dimension is defined and examples of its application to K=3, K=4 and K=7 central place systems are given. In order to apply the extended dimension to the K=7 central place system, which ordinarily lacks self-similarity, a procedure is applied to generate a K=7 system which does possess self-similarity. In Section 3, the hierarchical structures of higher order central place systems are illustrated and the fractal dimensions of these systems are obtained. Finally, conclusions and implications for research are given in Section 4.

2. The Fractal Dimensions of K=3, K=4 and K=7 Central Place Systems - and How to Make a K=7 Central Place System with Self-Similarity

One characteristic of Arlinghaus's central place model is that it permits many different sizes of market area, in contrast to the Christaller model, in which market area sizes are restricted to only three types: K=3, K=4 and K=7. In addition, Arlinghaus's model employs an upward perspective, in common with Lösch's model, while Christaller employs a downward perspective, a distinction of some importance as Parr (1983) makes clear. In effect, Arlinghaus's model can be considered a hybrid of the Christaller and Losch models.

The difficulty confronted in generating a K=3 central place system using the method proposed by Arlinghaus can be explained as follows. Arlinghaus's central place system is built up from a fractal iteration sequence, in which a generator transforms a hexagonal initiator into a first teragon composed of three hexagons, to which the generator is again applied to yield a second teragon composed of three copies of the first teragon, and so on. Not only does the hierarchical structure of the central place system derived in this manner become very complex, but it is also very different from that described in the Christaller model. Indeed, it is difficult to explain the formation of such a complicated hierarchical structure in terms of economic and geographical processes. The hierarchical structure of Arlinghaus's K=4 central place system is even more complicated, and in either case, as the number of fractal iteration approaches infinity, the degree of complexity of the hierarchical structure increases rapidly. The idea that an urban system can be built up from such intricate hierarchical structures is very difficult for the economist or the geographer to accept.

As Arlinghaus points out, one of the difficulties in producing a hierarchical structure with a particular K-value is that it is not easy to identify the correct generator corresponding to this K-value. For example, it may be difficult to find the proper generator to form a K=9 central place system; indeed, Arlinghaus avoids discussing this particular system. Whether it is possible to find suitable generators to construct every possible K-value central place system remains to be seen, but should it prove impossible, then the procedure described by Arlinghaus would lose its claim to generality.

Another problem surrounds the values of the fractal dimensions for different K-value central place systems. In a sense, the fractal dimension is a quantitative expression of the degree of complexity of shape. It is reasonable to suppose, therefore, that a central place system with a larger K-value should possess a hierarchical structure with a more complicated shape. Arlinghaus's own conclusion, that the value of the fractal dimension of central place systems decreases as the K-value increases, is therefore inconsistent with our expectations.

The fundamental cause of these problems lies in the definition of the similarity dimension which has been adopted to generate central place systems and to calculate the fractal dimensions. The fractal dimension is a general term covering all dimensions which are able to take noninteger values (see Takayasu, 1986). A unified definition of dimension has yet to be agreed upon, however. The similarity dimension is but one of many kinds of fractal dimension, which is limited to the case of fractal figures with strict similarity. One of the remarkable features of a figure with strict similarity is its nested structure, which allows the whole structure of the figure to be perfectly reproduced in a subset of itself, a typical example being the Koch curve (see Mandelbrot, 1986).

Although a whole series of central place systems proposed thus far have been subject to the influence of the Christaller model, note that the K=3, K=4 and K=7 central place systems formed through the regular subdivision of a market area do not have nested hierarchical structures. Consider a concrete example: a K=3 central place system with hierarchical level m. If a K=3 central place system has a perfectly nested structure, the number of hexagonal market areas at level i-1 that are contained within a market area at level i is three, where i=1,2, ... ,m. However only one out of these areas is completely included within a market at level i, the center of which is also a center at level i. As for the other six market areas at level i-1, one third of each market area is included in the market area at level i, and the two thirds remaining are allocated to two adjoining market areas at level i. If the K=3 central place system had a perfectly nested structure, there would be three hexagonal market areas at level i-1 within a market area at level i. The same is true for the other two central place systems (K=4 and K=7) in the Christaller model. If the K=4 and K=7 central place systems have hierarchies with nested structures and self-similarity, they should include four and seven smaller hexagonal market areas at level i-1 within a market area at level i respectively. Although the central place systems generated by Arlinghaus through a

76

process of fractional iteration exhibit nested structures with self-similarity, they have complicated hierarchical structures which are very different from those described in the Christaller model.

In the following discussion, several definitions of the fractal dimension are given and an extended capacity dimension is proposed. The fractal dimensions of K=3 and K=4 central place systems are then calculated through the use of this extended capacity dimension in parts 1) and 2), while a modified K=7 central place system is described in part 3) and its fractal dimension obtained.

The capacity dimension is defined as follows:

$$D_c = - \lim \frac{\log N(e)}{\log e} \tag{1}$$

where $N(e)$ is the minimum number of d-dimensional spheres with radius e with which a bounded set in Euclidean space R^d can be covered (see Takayasu, 1986). The similarity dimension, on the other hand, is defined as:

$$D_s = - \frac{\log b}{\log s} \tag{2}$$

where b is the number of similar figures made from and included in an original figure at a reduced scale s ($0 < s < 1$). This definition is identical to the one given by Arlinghaus.

The capacity dimension is generally defined for application to empirical studies as follows:

$$D_r = - \frac{\log N(r)}{\log r} \tag{3}$$

where r ($0 < r < 1$) is a scale to measure the fractal figure and $N(r)$ expresses the number of elementary figures, such as spheres and squares, covering the fractal figure. The formula (3) shows that the relationship between $\log N(r)$ and $\log r$ would appear as a linear function with slope D_r on a logarithmic graph. If $N(r)$ is a continuous function and the fractal dimension D_r is defined as the slope at ($\log r$, $\log N(r)$), then:

$$D_r = - \frac{d\log N(r)}{d\log r} \tag{4}$$

where it is worth noting that D_r is dependent on r, which is not constant. The formula (4) is an extended fractal dimension (see Takayasu, 1986, for further discussion).

1) The K=3 central place system

The following notation will be utilized:

i = hierarchical level within the system ($i = 1,2, ... ,m$);
M_i = hexagonal market area at level i;
S_{M_i} = value of a hexagonal market area at level i;
c_i = central place of level i;
d_i = length of a side of M_i;
r_i = radius of a circle circumscribed within M_i;
C_{r_i} = circumscribed circle with r_i of M_i

Figure 1 shows a K=3 central place system according to the Christaller model. As previously stated, there is one M_{m-1} market area properly included and another six M_{m-1} market areas partially included within the M_m market area. This process continues from level m to level 1 in a recursive manner. It is possible to continue the process infinitely, if a central place system has a hierarchical structure with an infinite number of levels.

Let us suppose that $d_m = a$, the length of a side of M_m. If so, then

$$d_{i-1} = \frac{1}{\sqrt{3}} d_i = (\frac{1}{\sqrt{3}})^{m-(i-1)} a \tag{5}$$

is easily obtained.

Obviously,

$$S_{M_{i-1}} = \frac{1}{\sqrt{3}} S_{M_{i-1}} = (\frac{1}{\sqrt{3}})^{m-(i-1)} S_{M_m} \tag{6}$$

and

$$S_{M_m} = \frac{3\sqrt{3}}{2} a^2$$

follows.

If D_s is employed as suggested by Arlinghaus, then the fractal dimension of the K=3 central place system in the Christaller model is obtained as

$$D_s = -\frac{\log 3}{\log\left(\frac{1}{\sqrt{3}}\right)} = 2.0$$

from equations (5) and (6). Although Arlinghaus gives the fractal dimension of a K=3 central place system as about 1.26, her central place system is rather different from the K=3 system in the Christaller model. The latter has no nested structure, and thus it is not appropriate to use D_s. Instead of D_s, we can extend the definition of the capacity dimension D_c and apply it to the K=3 system to obtain a revised fractal dimension. Though D_r is an extended definition of D_c, we can remove the restriction that D_r is a constant for application to cases such as the K=3 system, such that

$$D(r_i) = -\frac{\log N(r_i)}{\log r_i} \tag{7}$$

is newly defined.

Next, let us consider C_{r_i}, a circumscribed circle with a radius r_i of M_i, which we can use to compute the fractal dimension of a K=3 system. Since $r_i = d_i$ through all the levels, $r_i = (1/\sqrt{3})a^{m-i}$. How many circumscribed circles are needed to cover M_m as efficiently as possible? It is generally rather difficult to find the minimum number of circles to cover M_m. If a rule of thumb is adopted that M_i is covered by a C_{r_i} centred at each c_i, then the number of C_{r_i} covering M_m exactly equals the number of c_i. The series of the number of c_i within M_i in a descending order of levels can be expressed as follows:

$$1, 2, 6, 18, 54, \ldots \ldots \tag{8}$$

It should be noted that a weight of either 1/3 or 1/2 is given to each c_i in the sequence described in equation (7) to calculate the number of c_i on either a vertex or a side of the hexagonal market area M_i.

If we now ignore the weights of 1/3 or 1/2 and take account of the fact that central place systems of the Christaller type possess successively inclusive hierarchies, then the sequence described in (8) can be rewritten as follows:

$$1, 7, 13, 37, \ldots \ldots \tag{9}$$

Since the number and radius of C_{r_i} are equal to the number of c_i and the length of a side of M_i respectively, it is sufficient for a C_{r_i} centred at c_i to cover M_i. Using equation (7), which depends on r_i, the fractal dimensions of M_i are each defined as $D(r_{m-1}) = -\log 7/\log(1/\sqrt{3}) = 3.5425$, $D(r_{m-2}) = -\log 13/\log(1/\sqrt{3})^2 = 2.3347$, $D(r_{m-3}) = -\log 37/\log(1/\sqrt{3})^3 = 2.1912$ and so on at each hierarchical level. As r_i approaches zero, $D(r_i)$ approaches 2.0 asymptotically, as shown in Figure 4. This means that as the numbers and areas of M_i embedded in M_m increase, so d_i approaches zero and at the limit M_m acquires a nested structure. The sequence $D(r_i)$, therefore, can also be considered an index of the degree of convergence between the hierarchical structure of a K=3 central place system and a pure nested structure.

2) The K=4 central place system

Following on from the K=3 central place system, the relationship between d_i and d_{i-1} is easily obtained as:

$$d_{i-1} = (\frac{1}{2}) d_{i-1} = (\frac{1}{2})^m \bar{d}_m^{(i-1)} = (\frac{1}{2})^m \bar{a}^{(i-1)} \tag{10}$$

so that

$$S_{M_{i-1}} = (\frac{1}{4}) S_{M_i} = (\frac{1}{4})^m \bar{S}_{M_m}^{(i-1)} \tag{11}$$

and $S_M = (3/\sqrt{3}/4)a^2$ given that a is the length of a side of a market area M_m in a K=4 central place system.

The series of frequencies of c_i in a K=4 central place system arranged in descending order of levels

$$1, 3, 12, 48, \ldots \ldots \tag{12}$$

is shown in Figure 2.

If, as with the K=3 central place system, we now ignore the weight of 1/2 and take account of the successively inclusive hierarchy in the central place system, then the sequence described in (11) can be rewritten as:

$$1, 7, 19, 73, \ldots \ldots \tag{13}$$

Hence, the sequence $D(r_i)$, the fractal dimension of the K=4 central place system, is given as $D(r_{m-1}) = - \log 7/\log (1/2) = 2.8074$, $D(r_{m-2}) = - \log 19/\log (1/2)^2 = 2.1240$, $D(r_{m-3}) = - \log 73/\log (1/2)^3 = 2.0633$ and so on. Figure 4 shows that as r approaches zero, so $D(r_i)$ in the K=4 central place system again approaches 2.0. Since $D(r_i)$ is an index of the degree of convergence to a pure nested structure at level i, it can be seen that the K=4 central place system has a better nested structure than the K=3 central place system through every level i, and approaches 2.0 more rapidly than the K=3 system.

3) The K=7 central place system

The key difference between the K=7 central place system and the K=3 and K=4 systems is that the market area at level m is not hexagonal. Its structure is a honeycomb composed of several hexagonal market areas of level m-1, which Christaller (1968) derived from the administrative principle. This means that there is no relationship of self-similarity between market areas at levels M_m and M_{m-1}, because areas at M_{m-1} should also have a honeycomb structure if M_{m-1} is similar to M_m.

In order to generate self-similarity between M_i and M_{i-1} through all levels in a K=7 central place system we must first revise the K=7 system in the Christaller model and consider instead a system such as that illustrated in Figure 3, in which M_i is composed of a honeycomb structure of six hexagonal market areas M_{i-1} and six empty rhombus-shaped areas through all hierarchical levels i.

The system shown in Figure 3 is constructed as follows. A hexagonal market area M_{m-1} is divided into six regular triangles which are each formed from an edge of the M_{m-1} market area and the two lines which join the center of the M_{m-1} area to the two adjoining vertices. Six central places c_{m-2} are then located at the centers of gravity of the six regular triangles, such that these six c_{m-2} centers and the one c_{m-1} center are also the centers of seven hexagonal market areas M_{m-2}, as shown in Figure 3. It is possible to continue this process indefinitely. The K=7 central place system thus obtained has not only has a hierarchical structure with self-similarity, but that structure is also nested. The inside of an imaginary hexagonal market area M_m at level m (surrounded by a thin continuous line in Figure 3) is composed of a honeycomb structure of seven hexagonal market areas M_{m-1} and six empty rhombus-shaped areas. M_m, then, has the same structure as any other market area M_i.

The sequence of centers c_i is

$$1, 7, 7^2, 7^3, \ldots \ldots$$

(14)

and the relationship between d and d_{i-1} is expressed as

$$d_{i-1} = \frac{1}{3} d_i = (\frac{1}{3})^{m-(i-1)} d_m = (\frac{1}{3})^{m-(i-1)} a \qquad (15)$$

where it is assumed that $d_m = a$ is the length of the side of the imaginary hexagonal market area. If a true market area at level i is supposed not to be a hexagonal market area M_i, but rather a market area with the honeycomb structure which remains after taking away the six empty areas, then the fractal dimenson of the system for any i can be derived from equations from (14) and (15), as follows:

$$D(r_i) = -\frac{\log 7^i}{\log (\frac{1}{3})^i} = 1.7712 \qquad (16)$$

3. The Fractal Dimensions of Higher Order Central Place Systems

Expanded central place systems of the Christaller type derived from the generating function $K = f(x,y) = x^2 + xy + y^2$, and with self-similarity through all levels (like the hierarchical structures of the K=3, K=4 and K=7 types), seem to be limited to just five additional types: K=9, K=12, K=13, K=16 and K=19. The procedure for constructing these higher order central place systems is basically the same as that already described for the K=7 system. The higher order systems are shown in Figures 5 and 6, though only the hierarchical relationships between any M_{m-1} and M_m are indicated. The relationship between any M_i and M_{i-1} can easily be established through iteration, in the manner illustrated in these Figures.

Let us use K_n, in preference to K=n, and define a new notation K^*_n to express a central place system that includes all central place and market areas except for the c_m central place. K_9 can then be represented as a central place system obtained by combining K^*_3, K^*_7, and c_m. Table 1 shows how to combine K^*_3, K^*_4, K^*_7 and c_m to form the K_9 and K_{19} systems. It can easily be seen from Table 1 that the structures of the central place systems above K_7 are composed of combinations of K^*_7, c_m and at least one out of the series K^*_3, K^*_4 and K^*_7. In addition to K^*_7, if either or both of K^*_3 and K^*_4 are included in a K_n central place system, then that system will lack a nested structure though it will have a hierarchical structure with self-similarity. K_n central place systems which consist only of two or more K^*_7 and c_m, however, will have self-similarity and nested structures.

The fractal dimensions of each K_n can also be divided into two types: those such as K_3 or K_4, which approach 2.0 asymptotically as r_i approaches zero, and the K_7

type, which remains constant and independent of a decreasing r_i. An example of the former is the K_9 central place system, for which the ratio of the length of a side of M_i to that of a side of M_{i-1} is given by

$$d_{i-1} = \frac{1}{3} d_i = (\frac{1}{3})^{m-(i-1)} d_m \qquad (17)$$

The fractal dimension of the K_9 system is $D(r_{m-1}) = - \log 13/\log (1/\sqrt{3}) = 2.3347$ at level m-1 and $D(r_{m-2}) = - \log 91/\log (1/\sqrt{3})^2 = 2.0530$ at level m-2. As the radius of the circle C_{r_i} that covers M_m approaches zero, so $D(r_i)$ approaches 2.0. K_{12} and K_{19} behave in a similar fashion, and their fractal dimensions are shown in Table 2.

An example of the latter type is the K_{13} central place system, for which the hexagonal market area at level i is composed of 13 hexagonal market areas at level i-1 and six frustum-shaped empty areas. If we imagine a hexagonal market area M_m as indicated in the K_7 central place system, then the ratio of the length of a side of M_m to that of a side of M_{m-1} is expressed as follows:

$$d_{i-1} = \frac{1}{4} d_i = (\frac{1}{4})^{m-(i-1)} d_m \qquad (18)$$

The series of the number of centers c_1 obtained as i decreases is:

$$1, 13, 13^2, 13^3, \ldots \ldots \qquad (19)$$

The fractal dimension of the K_{13} central place system for any i is given by:

$$D(r_i) = - \frac{\log 13^i}{\log (\frac{1}{4})^i} = 1.8502 \qquad (20)$$

In the same way, the fractal dimension for K_{19} can easily be obtained:

$$D(r_i) = - \frac{\log 19^i}{\log (\frac{1}{5})^i} = 1.8295 \qquad (21)$$

4. Conclusion

In this paper an alternative method based on the capacity dimension has been proposed to obtain the fractal dimension of a K_n central place system of the Christaller type, in contrast to the method based on the similarity dimension proposed by Arlinghaus (1985). The central place systems created using Arlinghaus's method, in which a generator acts on a teragon in an iterative manner, have complicated structures which are difficult to accept, particularly the structures for the K_3 and K_4 systems, which are exceedingly irregular in comparison with those described in the Christaller model.

The Christaller model has achieved broad acceptance, in spite of the fact that it is restricted into just three types (K_3, K_4 and K_7). Hence, a procedure to expand central place systems of the Christaller type within the range K_3 to K_{19} has been proposed in this paper, with each system corresponding to the numbers derived from the K-value generating function. These central place systems are grouped into two types: those such as K_3 and K_4 which do not have nested structures but have self-similarity, and the others such as K_7 which have both. The fractal dimensions of these systems have been obtained using an extended definition of the capacity dimension (equation 7). The results demonstrate that in the former group, the fractal dimension decreases towards 2.0 asymptotically as the radius of a circumscribed circle ^Cr_i, which covers a market area at level m, is increased, while in the latter group the fractal dimension is constant and smaller than 2.0 irrespective of decreases in the radius of circumscribed circles. Thus it appears that the extended definition (7) is useful in analysing any fractal figure which lacks a nested structure but possesses self-similarity, and as an index to express the extent to which such a figure approaches a nested structures. The geographic and economic implications of the fractal dimensions of central place systems remain to be determined, however, as Arlinghaus also points out.

Another interesting issue is the relationship between the so-called rank-size rule and fractal theory. The rank-size rule, one of the most widely accepted generalizations in quantitative geography, is generally thought to coincide with the Pareto distribution - which is also an important distribution in the theory of fractals. Batty and Longley (1986) have noted that "cities are clearly self-similar in a variety of ways", and the fact that cities of different sizes are subsumed in the abstract concept "city" seems to indicate that there are some important functions and activities working in common to maintain and develop cities irrespective of their size. Hence differences between cities may reflect little more than underlying differences in the sizes of these functions and activities. This seems to suggest in turn that there is a relationship between the rank-size rule and the fractal structures of these cities, since in mathematical terms fractal figures are just nested structures made up of a basic figure and self-similar but reduced scale clones of it obtained through a process of infinite recursion. Finally, one other interesting issue would be to explore the relationship between fractal theory and the general hierarchical model (Parr, 1978).

References

Arlinghaus, S.L., 1985, 'Fractals take a central place', *Geografiska Annaler*, 67B, pp.83-88.

Batty, M. and Longley, P.A., 1986, 'The fractal simulation of urban structure', *Environment and Planning*, A, 18, pp.1143-1179.

Christaller, W., 1969, *Location and Development of Cities* (in Japanese),Tokyo: Taimeido, (translated by J. Ezawa from *Die Zentralen Orte in Süddeutschland*, Jena, 1933).

Dacey, M.F., 1965, "The geometry of central place theory". *Geografiska Annaler*, 47B, pp.111-124.

Mandelbrot, B., 1986, *Fractal Geometry* (in Japanese). Tokyo: Nikkei Science (translated by H. Hironaka, from *The Fractal Geometry of Nature*, W H Freeman, 1983).

Parr, J.B., 1978, "Models of the central place system: a more general approach". *Urban Studies*, 15, pp.35-49.

Parr, J.B., 1983, "City size distributions and urban density functions: some interrelationships". *Journal of Regional Science*, 23, pp.283-307.

Takayasu, H., 1986, *Fractals* (in Japanese). Tokyo: Asakura Shoten.

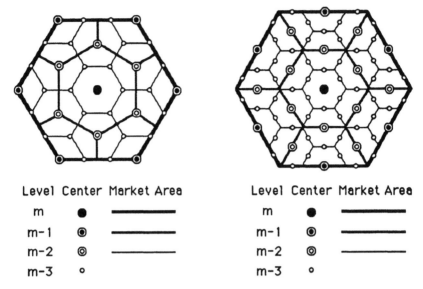

Figure 1: K=3 Central Place System   Figure 2: K=4 Central Place System

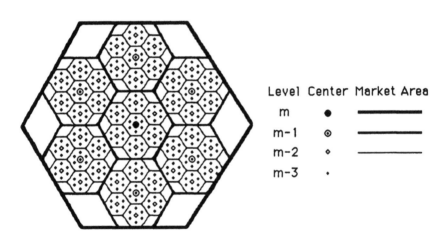

Figure 3: K=7 Central Place System

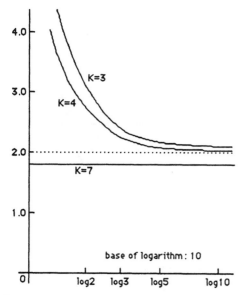

Figure 4: Fractal Dimensions of K=3, K=4
and K=7 Central Place Systems

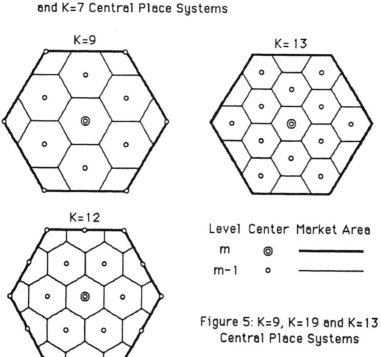

Figure 5: K=9, K=19 and K=13
Central Place Systems

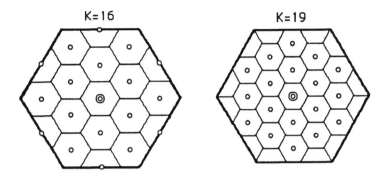

Level Center Market Area

m   ◎   ──────

m-1   o   ────

Figure 6: K=16 and K=19
Central Place Systems

Table 1: The formation of higher order central place systems by combining K^*_3, K^*_4, K^*_7 and c

K_n	Combination	K_n	Combination
K_9	K + K + c	K_{13}	K + K + c
K_{12}	K + K + K + c	K_{19}	K + K + K + c
K_{16}	K + K + K + c		

Table 2: Fractal dimensions of K_n central place systems

K_n	$D(r_{m-1})$	$\lim D(r_i)$ (as r_i approaches 0)
K_9	2.26	2.0
K_{12}	2.37	2.0
K_{13}	1.85	2.0
K_{16}	2.12	2.0
K_{19}	1.82	2.0

Comparative Research on the Hydrological Characteristics of some British and Japanese Drainage Basins

Akihisa Yoshikoshi
Nara University

1. Introduction

This paper attempts to clarify the hydrological characteristics of some British and Japanese basins on a comparative basis, using estimates of the total volume of water in each drainage area and of seasonal and yearly fluctuations in that volume. The method of research adopted is to compare the hydrological characteristics of the Exe and Nagara Basins, with particular reference to the water balance and flood conditions.

2. A general view of the two rivers

The drainage basin of the River Exe lies in the southwestern part of the British Isles. The Exe Basin is taken to include the major tributaries, the Rivers Creedy, Culm and Clyst, and has a drainage area of $1,462km^2$. The Exe rises on Exmoor at a point more than 450m above sea level, but within 8km of the Bristol Channel. Its total course is about 100km long. The study area consists of the upper part of the Exe Basin, and has a drainage area of $600.9km^2$.

The geological formation of the basin contains a variety of rock types, which in turn exhibit a wide diversity of lithology, ranging from the indurated slates and grits of the Devonian to the marls and sandstone of the Triassic. This geological diversity is paralleled by contrasts in relief and related hydrometeorological conditions. The annual average precipitation exceeds 1,035mm, with the wettest part of the basin to be found on Exmoor and the driest part in the Exe Estuary.

As for land use, most of the basin is occupied by crops or pasture, although some 8% is under woodland and another 3% or so is devoted to urban uses.

In many respects the Exe is typical of medium or small scale rivers in Great Britain. From the hydrogeomorphological point of view, it has a steep channel slope and a relatively high drainage density, while from a hydrometeorological perspective it has a large discharge volume and a high discharge ratio, a consequence of its location in the western part of the British Isles where there is abundant precipitation. The basin can also be considered typical of rural basins in Britain.

The River Nagara, on the other hand, is located in the central part of Japan, and is one of the *Kiso sansen* (three rivers of Kiso, Ibi and Nagara). The *Kiso sansen* follow separate courses now, but up to about a century ago they cut an erratic path across the Nôbi Plain as one river, which changed its course every time there was a flood. The Nagara, which issues from Mt. Dainichi-Dake (1,709m), flows through Gifu City before entering the Nôbi Plain. It has a drainage area of 1,985km^2 and a stream length of 158km. The study area consists of the upper part of the Nagara Basin, which has a drainage area of 1,066 km^2 and a stream length of 70km.

The geological formation of the basin consists of Pleistocene andesite in the north and west, where the channel slope is steep, and rhyolite rocks of the Palaeocene and Cretaceous in the west and south. The central part of the basin is occupied by Palaeozoic sandstones and slate. The annual average precipitation approaches 3,000mm, with the wettest part of the basin to be found in the north. As for land use, most of the basin is occupied by woodland and rice fields.

The Nagara can be considered a typical medium-scale Japanese river, from both a hydrogeomorphological and a hydrometeorological point of view. The Nagara Basin is also a typical rural basin, and human activities have had few effects within the study area.

Obviously then, these two rivers share a number of characteristics, which makes a comparison between them particularly appropriate.

3. A comparison of water balances

The water balance is a method of accounting for all water inputs (precipitation and so forth) and all water outputs (discharge, evapotranspiration etc.) within a certain area over a certain period. Calculation of the water balance allows us to determine the hydrological characteristics of a drainage basin.

The water balance equation can be expressed in simple terms as follows:

$$P - D - ET = \Delta S \qquad (1)$$

where
P = precipitation;
D = discharge, including the discharge of river water and groundwater;

ET = amount of evapotranspiration; and

ΔS = change in water storage, which over a period of many years would average zero.

The items included in the water balance equation for the Exe Basin are shown on Figure 1. Monthly precipitation has been calculated as the average for twenty precipitation-gauging stations located within the study area. Monthly discharge has been estimated from the daily mean flow at Thorverton, while monthly evapotranspiration represents the average for two evapotranspiration gauging stations, and has been obtained using the Penman method. The change in water storage ΔS was calculated from formula (1). Over the year as a whole ΔS is not zero, because of water withdrawal along the Grand Western Canal.

Figure 1 also shows the components of the water balance for the Nagara Basin. Monthly precipitation was calculated using the isohyetal line method, while monthly discharge was estimated using data on the daily mean flow at Mino City, Gifu Prefecture. Monthly evapotranspiration was estimated by the Thornthwaite method, and ΔS was calculated from formula (1).

The Exe Basin experiences more precipitation in winter than in summer, while the Nagara Basin has a summer rainy season. Total precipitation in the Exe Basin is under half that in the Nagara Basin. The discharge ratio for the Nagara Basin is about 83%, but that for the Exe Basin is only about 63%, a difference which is caused by the volume of precipitation. Changes in monthly evapotranspiration are similar between the two basins, as are the monthly ΔS figures. In both cases, water which has been stored in winter is consumed in summer.

4. A comparison of floods

The figures presented in the previous section describe the hydrological characteristics of these basins over the long term. Over much short periods, however, very different hydrological characteristics can be displayed. Typical of such characteristics are droughts and floods. Here we will consider the latter with reference to the two basins.

Floods in Britain can stem from a variety of causes, including heavy rain, snowmelt and high tides. Let us consider in detail the so-called Christmas Flood of 1985 in the Exe Basin. The Christmas Flood of 1985 was a fairly small scale event within the history of floods in Great Britain, but it can be thought of as typical of floods caused by heavy rain and high tides.

It started raining in the Exe Basin on 21 December, 1985, and eventually the river flooded an area of some $30km^2$. The peak flow of the floodwaters at Thorverton exceeded 200 m^3/sec.

On the other hand, the upper part of the Nagara Basin is rarely flooded, but the lower part is habitually affected. The latter area is known for its *waju* (*wa* means a ring and *ju* an area encircled by a ring levee). There are many *waju* along the lower reaches of the *Kiso sansen*, and this area contains the largest concentration of *waju* in Japan. There are many characteristic houses known as *mizuya* in this area. The *miyuza* is a kind of shelter built on a platform of stones or soil. It is usually built near the main house, and is used as a storehouse for food. If a flood arises, people can live in the *mizuya* until the water goes down.

Here I would like to examine one typical flood of the River Nagara, the September Flood of 1976. During this flood the peak flow at Mino was over 3,000m^3/sec, and the damage may be summarized as follows: 19 persons dead or injured, 38,000 buildings damaged, and $222km^2$ flooded.

The main differences in the flood characteristics of the two basins are shown in Table 1. The figures in parentheses are data for the Christmas Flood of 1985 and the September Flood of 1976. There is a clear difference in the discharge, which can be attributed in turn to the difference in the level of precipitation. Moreover, there is a marked disparity in the degree of damage caused, due to differences in the use of land near the river. In Great Britain, land use near rivers is regulated by law and is confined to pasture, arable and so on in a deliberate attempt to mitigate flood damage. Land use near the River Nagara, however, consists of a mixture of arable land, residential areas and industry, a situation quite typical of other Japanese rivers. There has also been excessive reliance on dikes - another reason for the more extensive flood damage in the Japanese case.

5. Conclusion

The results of this research reveal great differences between the hydrological characteristics of these two rivers, in spite of their spatial similarities. There are also some similarities, however, such as the changes in monthly ΔS. The next stage of this research will consist of a comparison of other hydrological characteristics and an investigation of their causes.

Table 1: Differences in Flood Characteristics

	Exe Basin		**Nagara Basin**
Total precipitation	Small amount	(105mm)	Large amount (1052mm)
Precipitation intensity	Weak	(46mm/day)	Strong (278mm/day)
Mean channel slope	Gentle	(4.5m/km)	Steep (10.8m/km)
Flood discharge (Daily mean)	Small amount	($152m^3$/sec)	Large amount ($3099m^3$/sec)
Discharge ratio	Low	(67%)	High (99%)
Effect of tide	Large		Small
Land use near river	Grassland		Arable land, Urban area
Dikes	None		High dikes
Damage	Small amount		Large amount

Figure 1: Water Balance Figures for the Exe Basin and the Nagara Basin

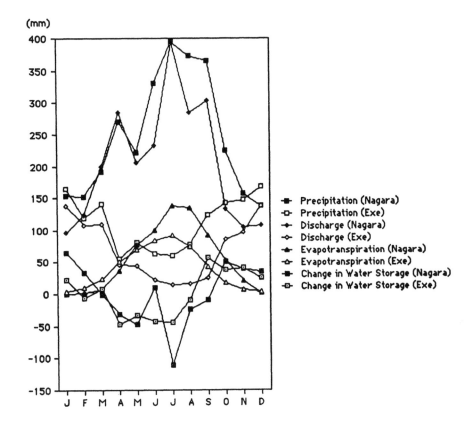

(mm)

- ■ — Precipitation (Nagara)
- -□- Precipitation (Exe)
- ◆ — Discharge (Nagara)
- -◇- Discharge (Exe)
- ▲ — Evapotranspiration (Nagara)
- -△- Evapotranspiration (Exe)
- ■ — Change in Water Storage (Nagara)
- -□- Change in Water Storage (Exe)

On the Consumption of Space: A Paradigm for Human Geography

Iwao Kamozawa

Hosei University

When in 1947 R.E. Dickinson denied the validity of applying the principles of the natural sciences to the interpretation of the phenomena treated in human geography, and A.K. Philbrick ten years later stressed even more strongly the role of human beings as the core factor in the formation of a region, it became clear that the ascendancy of traditional regional geography, which had been based on the principles of the natural sciences, had come to an end.

Although so-called traditional regional geography seems to be moribund, regional geography itself cannot necessarily be declared dead. Regional geography is and should be a viable part of an anthropocentric human geography. A number of tools for analysis have been developed by quantitative geography, but at the same time, it has not succeeded in formulating any general scientific laws, despite its original intent. This did not happen by chance: quantitative geography seemed to seek laws no different from those which underlie the natural sciences. Indeed, in this sense quantitative geography has much in common with Romanticism in its search for natural bases, as described by earlier Romantic geographers such as F. von Richthofen or F. Ratzel.

Clearly, we have to pursue a type of human geography other than one based on natural laws, and it seems reasonable to declare that the proper path for human geography is a human-oriented one (cf. D.K. Forbes's "The Geography of Underdevelopment", 1984). The problem remains, however, of how this idea can be applied to the analysis of real phenomena. In this paper I would like to suggest a solution.

The consumption of space

Each social group consumes the space it requires, and as a consequence it becomes identified with a region. This is my fundamental theme in identifying a human-oriented geography. We do not usually use the expression "consuming space". Space is not exhausted after consumption, unlike money for example, which is exhausted after being consumed. It seems improper to speak of consuming space, as it is by nature non-exhaustible. On the other hand though, we do talk about consuming time. Time is always present, along with space, in our real lives: thus we say, for example, that "now we are here". If we say time is consumed, why should we not also say that space is consumed? If our real existence is always conditioned by time, then it is also always conditioned by space. If we always consume time in our lives, then we also always consume space. Ultimately, each individual person consumes time and space, but we cannot hope to study every single individual if our purpose is to make human-oriented generalizations in the field of human geography, for this requires that to some extent we treat human beings as a mass. If we do treat individuals instead of masses, then we are apt to introduce subjective biases, as is often apparent in the field of humanistic geography. This is why I have used the expression "each social group consumes the space it requires" above.

The relationship between using space and consuming space can be explained as follows. When a social group comes in contact with a piece of space at a certain time, it is with the intention of using that piece of space. As soon as the social group begins to use the space, however it has no alternative but to create its own unique way of using the space, in the sense that the space was not pre-existing or unalterable, but newly created in the context of an emerging relation between society and nature. It may be better then for us to use the word "consumption" instead of "use", in order to mark the difference between space as a physical entity and space as a social entity, though its substance is ultimately defined by nature.

In keeping with the natures of human groups and of space, the same space will be consumed differently by different human groups: exploitation and protection of the same natural resources in a region by different social groups at the same time is one of the remarkable characteristics of life on earth which is universally observed. Hence, we can use the concept of the consumption of space by different social groups to analyse the balances, imbalances and conflicts within regions which in turn produce the ever changing regional configuration of our world.

Some applications of the concept

1) Population geography
The concept of 'ras-selenie', which has long been a feature of Soviet geography (cf. "Kratkaya Geograficheskaya Entsiklopediya", Vol. 3, 1962, Moskva, pp. 338-339), is valid whenever a geographer attempts to analyze both aspects of the population of a region, the static aspect (settlement) and the dynamic one (migration), because the concept combines the two. At present, however, the concept is confined to the field of population geography. Nevertheless, it can be incorporated as part of the concept of the consumption of space: some social groups consume space relatively statically

in their pattern of settlement, though they are always moving in their daily lives as commuters and so on, while other social groups consume space relatively dynamically when they are migrating.

2) Regionalization in daily life

The normal daily lives of working people are divided into two sections, not only by social forces but also by time and space: a home life section and a business life section. Each of the two sections requires a different site in most cases. In addition, the principles of behaviour required in each section are different: in a capitalist society for example, generally speaking one does not seek economic efficiency in one's home life, but one should seek it in one's business life. This underlies one of the basic conflicts in the lives of working people in contemporary capitalist societies. Daily movements from home to place of work (and the different uses of time in a day) by the majority of people is a factor in regionalization (region making): in other words, the competition involved in different types of space consumption (and time consumption) of a working person in a day functions to create a region.

3) Rural regionalization in the case of Japan

Rural society in Japan has changed and is still changing rapidly and fundamentally, mainly because of the rapid decline in the status of the agricultural sector in the national economy. For most rural households in Japan, agriculture is no longer the largest single source of income. The main income comes from work in construction, the services or the industrial sector. In winter, for example, many householders in the Tohoku district leave their villages to work at construction sites in the Keihin area. Other householders in the same district go to work in offices, teach in schools or work in factories. Those offices, schools and factories can be located either within the village or outside.

In such a situation, it makes little sense to focus solely on the agricultural sector or upon peasant farmers if one is seeking to understand the total regionalization process within rural society. The concept of space consumption becomes valid here, as this concept can cover a number of social groups occupying a region simultaneously, and remarkable contrasts can be observed in the types of space consumption undertaken by the various types of worker mentioned above. Indeed, there are various conflicts between the sectors in which the inhabitants of the region are working. For example, the more that local inhabitants engage in non-agricultural occupations, so the more the affect on the agricultural sector itself.

4) World regionalization

As a final example, the concept of the consumption of space can be applied to world regionalization, within which both leading groups and subordinate groups can be observed. The typology of space consumption between the two groups is clear, and it is also clear that there are various types of space consumption within each group.

The concept can be further applied, for example, to the problems of ecological systems. Space consumption can be of a system-preserving type or of an exploitative type. Struggles between these groups represent clear conflicts.

Part V: Regional Geography

South Asian Studies in Japan

Masanori Koga
Hitotsubashi University

Prior to World War II, South Asian Studies in Japan were confined almost exclusively to Indology, and in particular to the study of Buddhist philosophy and Buddhism. Several Imperial universities had chairs in Indology within departments of Philosophy or Religious Studies, while two colleges of foreign languages taught Hindi and Urdu. In almost every university there was a department of Oriental History, but this consisted of Chinese History and nothing else. It was only on the eve of World War II that South Asian Studies were introduced into the social sciences. Most of this work was undertaken outside the academic world, however, in semi-governmental organizations such as the Tôa Institute (Tôa Kenkyûjo), the research section of the Manchuria Railway Co. (Mantetsu Chôsa-bu) and the Indian Research Unit (Indo Sôgô Kenkyû Shitsu). Needless to say, these organizations were designed to contribute to the government's expansionist policies. It was a historical irony, however, that many of the researchers working in these institutions were actually converts from Marxism.

At that time most of the works available in Japanese were translations of British publications, and it is no exaggeration to say that these formed the starting point of South Asian Studies after the War, for the reason that up to the middle of the 1960s, Japanese scholars had almost no access to foreign publications, due to severe constraints on the availability of foreign exchange. For example, in order to visit India for research purposes, it was necessary to enroll as a student under an Indian Government Scholarship. During the 1950s the first groups of Japanese students began to study at Indian universities, and they subsequently became the pioneers of South Asian Studies within the social sciences in Japan. By the late 1950s, the Institute of Developing Economies - a semi-govermental institution - had also been established.

Since then, two colleges of foreign languages have been established, the first in Tokyo and later, the University of Foreign Studies in Osaka. These, together with the Institute of Oriental Culture, (which is now attached to Tokyo University), form the major centers of South Asian Studies in Japan.

Generally speaking, the Institute of Developing Economies specializes in contemporary political, economic and social studies, the Institute of Oriental Culture specializes in historical studies, and the universities of foreign affairs specialize in Indian/Pakistani literature and historical studies. As far as geography is concerned, it was the Department of Geography at Hitotsubashi University which took the initiative in South Asian Studies.

Up to the 1960s, the most popular areas of study were the nationalist movement, the rural economy, and land tenure systems in India. As the number of scholars has grown, however, the coverage and variety within South Asian Studies has also expanded, from archaeology to contemporary history, and from folklore to political science.

The main characteristics observable in recent studies are the increasing level of specialization by region and by topic, as well as diversification. An increasing number of scholars have mastered not only Hindi, but also Bengali, Tamil, Malayalam, Marathi, Gujarati and Punjabi, and are specialists in the respective states where these languages are spoken.

In the past most research was concerned with economic history, including the analysis of contemporary political and economic processes, but today many other fields of research are also being explored. These include defence, the caste system - particularly the problem of the 'untouchables' - communalism and urban problems.

Almost every year since 1963, a seminar for South Asian Studies has been held on an ad hoc basis in the summer vacation, encompassing a wide range of subjects and topics. This seminar is open to everyone interested in any aspect of South Asian affairs: not only professional scholars and researchers, but also journalists, high school teachers, specialists in various aspects of Indian culture, and even employees of commercial firms. This seminar has played an important role in stimulating interest in, and diffusing knowledge about India. It has offered students a good opportunity to familiarize themselves not only with the current situation in South Asian Studies, but also with other specialists in each subject. The numbers in attendance usually range from 150 to 200.

Apart from this seminar at the national level, there are four other major regional organizations for South Asian Studies. The longest standing of these is in Tokyo, while the others are in Osaka, Nagoya and Fukuoka, and all have regular monthly or bi-monthly meetings. However, these are not necessarily appropriate for scholarly discussion in depth. Therefore, some scholars have organized themselves into small study groups appropriate to their own specific fields, eg. the Bengal Economic

History Study Group, the Indian Archaeological Study Group, the Hindi Literature Study Group, etc. Some of these study groups issue newsletters or bulletins.

Since the 1970s, scholars in many other area studies fields have organized their own associations - the Association for Southeast Asian Studies and the Association for African Studies are examples. This has in turn given an important impetus to establishing an Association of South Asian Studies.

Three other factors should also be mentioned. First, as the number of scholars specializing in South Asian Studies grows, so each tends to become more and more narrowly specialized in a specific aspect, a specific area and a specific period. Because of this, some scholars have expressed increasing concern over the need to exchange ideas about the present state and future direction of research work across several disciplines, and to create a common forum.

Second, the need has increasingly been felt for a representative body for the promotion of research work and the solution of problems which most scholars have been facing in common. For example, the recent tightening of regulations concerning research visas in India has increased the difficulties experienced by many scholars who are trying to pursue field surveys or studies based on documentary evidence in India. Consequently, a demand has arisen to station some scholars in India permanently as academic liaison officers. If a recognized Association is established, it should become an effective means of fulfilling this demand.

Third, it has now become common practice among South Asia specialists in Japan to receive foreign scholars from various countries. So far, these foreign scholars have been hosted by individual Japanese scholars or institutions; however, it must be recognized that a more appropriate body for this purpose would be a comprehensive organization.

The date for the inauguration session of the Japanese Association for South Asian Studies has been set as the beginning of October 1988, marking a two year interval since the launching of its British counterpart.

From the Walled Inner City to the Urban Periphery: Changing Phases of Residential Separation in Damascus

Masanori Naito

Department of Social Geography
Hitotsubashi University

Damascus, as well as the other ancient cities in the <u>Bilad al-Sham</u> (Greater Syria), is known as a crossroads of Islamic and Christian civilizations. Rather than these cultures blending, however, a dichotomy has emerged between the Muslims and the Christians. Although residential separation between Muslims and Christians in the Old City (the area inside the ancient city wall) dates from the Arab conquest of Damascus in the 7th Century, it assumed a new dimension from the mid-19th century, as the Ottoman provinces gave way to European colonialism. Under the French religious protectorate the Christian communities became spatially segregated from the surrounding Muslim majority, and remained so even after the withdrawal of foreign troops in 1946.

Since Independence the dominance of the Syrian capital has led to a rapid concentration of population. With the influx of rural migrants (since the early 1960s) and of Palestinian refugees, several settlements housing religious and ethno-regional minority groups have developed in the peripheral areas of the city. As a result, new forms of residential separation have emerged. The exclusive settlements of the Druzes and 'Alawis are particularly important, for the improvement in the political status of these groups has created a degree of tension with the Sunni Muslim majority.

1. Christian-Muslim Separation Re-examined

Prior to Independence in 1946 there were several settlements structured along religious or ethnic lines. The oldest surviving religious communities are those of the Jews and the Christians, located at the eastern end of the Old City. The area from Bab Tuma southward to al-Bab al-Sharqi is the largest Christian quarter in Damascus, and within it there is little residential separation between the various sects, such as Greek Orthodox, Greek Catholic, Roman Catholic, Syriac, Armenian Orthodox and Catholic, and Maronite Christians.

In the 19th century the Christians were but a small religious minority in Damascus. From the mid-19th Century to the Mandate period in the 1920s, however, the

Christians often faced the threat of clashes with the Sunni majority. The first such incident broke out in Bab Tuma on July 9, 1860, when several thousand Christians were murdered and their residences attacked and burned. The mobs consisted mainly of urbanite Sunni shopkeepers, craftsmen and peasants, but scores of Sunni Muslim notables were also involved. This incident was triggered by resentment amongst the urbanite Muslims, especially those who had been suffering serious losses in the bazaars as a consequence of the Christians' privileges in external trade. Nevertheless, thousands of the Christians were sheltered by Muslim merchants and notable families such as the Jaza'iris, while there was also serious factional infighting amongst the Muslim notables themselves.

The riots in Damascus and the civil war in Lebanon provided a pretext for French intervention in the <u>Bilad al-Sham</u> in order to protect the Christians. The French drew up a blueprint for "segregating the Christians", and despite the subsequent return of civil order this policy persisted right through to Independence.

In 1920 the French Mandate was finally authorized by the European powers at the San Remo Conference. At that time the economic circumstances of the Christian community had deteriorated because of the effect of World War I on trade with Europe. For this reason the Mandate should have been welcomed by the comprador bourgeoisie, yet the Christians' attitude during the Mandate was cautious.

The Great Revolt, initiated by the Druze uprisings in the southern provinces, reached Damascus and its surrounding villages in 1925, by which time it had become a popular, nationalistic, and anti-imperialist rebellion. Once again the Christians in Damascus faced the threat of riots, yet they refrained from any overt political action, preferring instead to take a "wait and see" attitude.

Perhaps the Christians feared a recurrence of the July 1860 riots. Their passive attitude can also be attributed to conflicting class interests, however. The merchant-moneylending class stood to lose from the financial ruin of their Muslim customers, and thus desired a quick end to the Revolt, but craftsmen and peasants were largely uninterested, since their political and economic status was far below that of the Europeans' local agents. At the same time an external factor was at work - intermediation by a notable family, the Jaza'iris, on the Christians' behalf. The outcome was that while some families in exposed locations took refuge in the Bab Tuma quarter, the Christian community as a whole escaped serious injury.

The Christians' behaviour seems to reflect an awareness of their political situation; their survival was in the hands not of the French but of the notables and of the new nationalist forces. Their privileged status as the Europeans' agents offered little security, and by adopting a neutral stance to the Revolt the Christians could distance themselves from the policy of segregation pursued by the French.

Since Independence, the Syrian government's stance on the Christian communities has been a reflection of its own secularism. While the secularization policy created some dissatisfaction amongst the largest section of the population (which is Muslim), it was a practical necessity if national integration was to be achieved.

The exclusivity of the Christians' residential quarter in the Old City has persisted, but newly married couples often take the opportunity to move northward outside the wall to Qassa'a, the old European residential area. This relocation is due in part to inferior housing conditions in the Old City, but it also indicates a degree of modernization and westernization among Christian nuclear families. Gradually, the Muslims living in these new residential areas have become intermixing with the Christian inhabitants, with the result that a specific Christan quarter cannot be identified.

Meanwhile, the elderly Christian inhabitants remaining in the Old City have been renting out any vacant rooms to Christian in-migrants, with the result that the inner-city Christian community has been able to sustain its population with newcomers of the same faith. Socio-religious strictures have prevented the Muslims from following suit, however, with the result that the Muslim population within the Old City has declined.

The residential separation of the Christians may therefore be interpreted simply as a cultural phase, without implying any specific discrimination by the Muslim majority. This peaceful situation is guaranteed however only as long as they refrain from political involvement. The "wait and see" attitude has a very pragmatic basis in the present political climate.

2. The Emergence of Residential Separation on the Urban Periphery

Meanwhile, the emergence of various settlements on the urban periphery has created a new source of political unrest. These settlements are exclusive in the sense that each community consists of the same ethnic or religious minority. Their growth dates from the 1960s, when large numbers flooded into Damascus and its outskirts.

The exodus of rural 'Alawis, Druzes and Kurds to Damascus occurred in response to severe regional economic inequalities. The arrival of Druzes and 'Alawis was also linked, however, to their deepening involvement in political activities after Independence, while the Kurdish migration followed the severe drought in the Jazira region in 1983/84.

In contrast, the concentration of Palestinians and Shi'ites into Damascus was caused by external factors linked to Syria's foreign policy. The establishment of Israel in 1948, and its invasion of the Golan Heights in 1967, resulted in the exodus of more than two million Palestinians, of whom 10 percent were accepted by Syria as refugees. By 1984 more than 170,000 Palestinian refugees were being sheltered in camps scattered through Damascus and its surroundings. The Shi'ite community consists of various ethnic groups, such as settled Iranian pilgrims and Lebanese refugees from the 1982 Israeli attacks on Southern Lebanon.

Druze migrants have settled in the three villages in which they have traditionally formed a firm majority. Most had previously lived in the the Jabal Druze (Druze Mountains) in the Hawran region. In 1955 the population of Jaramana (the largest Druze settlement on the outskirts of Damascus) was 4,085, but by 1981 it exceeded 65,000, of whom 60 percent were Druzes. The total number of Druzes inhabiting the three villages was about 50,000 in the early 1980s, almost one third of the entire Druze population in Syria. In the late 1960s, however, the government began to restrict the concentration of Druzes in Jaramana by constructing a large refugee camp for 25,000 Palestinians nearby.

Political and economic conditions in Damascus since the 1960s accelerated the rural-to-urban drift of the Druzes. Druze military officers gained power within the Ba'th party through the 8 March Ba'th Revolution in 1963, and despite their defeat in factional infighting with the 'Alawis in 1966 they still maintain a degree of influence both in the army and in government circles. Subsequently the growth of the Syrian economy in the first half of the 1970s, which augmented industrial investment in the Damascus region, proved an additional strong attraction for poverty-stricken peasants such as the Druzes and 'Alawis.

As for the 'Alawis, they have generally been considered the largest Muslim minority sect in Syria. Their population is currently estimated to be around one million (10 percent of the national total), and they have been moving into Damascus from the mountain range behind Lataqiya, where they had previously suffered from economic subordination to the Sunnis, as well as social discrimination as a mountain-dwelling heterodox minority. In was within this broader inter-regional context that the 'Alawis became spatially segregated from the Sunnis in the urban environment as well.

After Independence many young 'Alawis were drawn to the army in the hope of improving their economic and social status, and they gradually expanded their political influence within the army through the Military Committee of the Ba'th party. A Ba'th government was established by the 8 March Revolution in 1963, and 'Alawi army officers came to power through a sequence of military coup d'etats lasting through 1970, when General Hafiz al-Asad expelled his intra-communal rivals, to established himself as President in the following year.

From then on, the 'Alawis flooded into both Lataqiya and Damascus, eagerly expecting a share of the privileges. Migrants to Damascus became concentrated in the al-Mezze Jabal quarter, which used to be their military base. The formation of this exclusive settlement did not simply reflect strong communal affinity however, it was also needed for security.

A series of socialistic measures were enforced under the first Ba'th regime in the 1960s, despite resistance from the urbanite Sunni bourgeoisie in particular, in order to facilitate national integration. One consequence of the rise of the minority sects, however, has been that the earlier consensus on socialistic policies has gradually been eroded. For the predominant Sunnis the Ba'th regime represents little more than sectarianism at their expense, while the harsh suppression of the Muslim Brotherhood uprisings in the early 1980s made the 'Alawi community the target of nationwide Sunni resentment.

In the early 1980s a number of 'Alawi party members were murdered, and the 'Alawis separated themselves off from the majority Sunnis in response. By and large they have not resettled outside of the initial settlement in al-Mezze, with the result that a continuous in-migration of 'Alawis has been required to maintain their dominance and security.

At Independence, the Syrian government was faced with the dual challenge of achieving national integration in Syria's pluralistic society and of curtailing the political, economic, and social privileges of the urban notables. National integration was greatly hindered, however, by the division of the country into various local territories which had initially been fostered by the French, such as the "états" established by the 'Alawis and Druzes under the French Mandate. Sectarianism among the minority groups failed to emerge during the first stage of state building from 1946 to 1963, although regional imbalances along ethnic and religious lines persisted. Throughout this period, the government focussed its efforts upon the expulsion of urban notables. After the renunciation of the Union with Egypt in 1961 however a succession of conflicts threw the state into serious political confusion, thereby worsening its economic position.

When the Ba'th party gained power in 1963, the initial phase of national integration was already in serious trouble. Alongside its socialistic measures, the Ba'th government also fostered the economic and political centralization of the administrative system. As a result, local autonomy became severely restricted. The 1969 Provisional Constitution defined the Ba'th party as the leader of society and the state, but deleted all reference to freedom of assembly and association. Subsequent uprisings by the Muslim Brotherhood in Aleppo and Hama were vigorously put down by the army, an action that awakened the people of these two cities to the risks associated with any anti-government actions or regionalist movements.

Meanwhile, the fundamental character and composition of the Ba'th party was itself undergoing change. The secular and socialistic ideologies of the Ba'th proved attractive to minority sects such as the 'Alawis and Druzes, while their rise within the party undoubtedly strengthened its sectarian tendencies. As a result, the national integration of Syria has been implemented through the 'Alawi's sectarian leadership of the Ba'th party, despite the gradual weakening of national cohesion. The centralizing measures enforced by the Ba'th regime have caused rapid growth in the population of Damascus since the 1960s, while the party's sectarian tendencies have fostered the formation of exclusive settlements by minority groups.

3. Future Prospects

Since the late Ottoman Period in the 19th century, various inter-communal tensions have lain at the heart of Syrian political life, and have given rise to residential separation between religious and ethnic minorities. Residential separation of the minorities in the city of Damascus was not a result of discrimination by the Sunni Arab majority, however; social discrimination linked to religious or cultural differences became salient only after the minority had gained dominant positions and privileges. In this sense, the residential separation of the 'Alawis and Druzes can be seen as a "contra-segregation" of a privileged minority from the Sunni majority, although this does not imply any form of socio-cultural discrimination against the latter.

Will this separation eventually dissolve, with the reduction of tension between the 'Alawi and the Sunni majority? Many doubt whether this can be achieved at all easily. The government has made some positive gestures, such as the amnesty for members of the Muslim Brotherhood in 1985 and the conversion of the Metropolitan Defence Guards' post in the al-Mezze quarter next door to the 'Alawis settlement into a public park. Yet in part these measures simply reflect the state's acute economic problems. The balance of payments has deteriorated since the early 1980s with the decrease in oil revenues, the withdrawal of foreign aid, and increasing military involvement in Lebanon. As a result, the government cannot afford to disburse funds to maintain internal security. And even assuming that an 'Alawi-Sunni collective leadership is established in the future, it is still not certain that it will be able to reduce other inter-communal tensions.

By the mid-1980s, the 'Alawis leadership faced a situation in which the agglomeration of minorities in exclusive settlements was leading to increased tensions between minority communities and the urbanite Sunni majority, a form of internal division that facilitated its ability to govern from a minority base. Hence, unlimited rural to urban migration has been accepted without any substantial restrictions. In this sense, the agglomeration process can be seen as a spatial strategy fostered by the state to protect its security. It should be also stressed, however, that each minority group has its own political stance with respect to the present Ba'th regime. Hence, the settlements on the urban periphery may yet be considered a double-edged sword by the Ba'th government.

Section 2: British Perspectives

The Place of Japanese Studies in British Geography

John Sargent

School of Oriental and African Studies
University of London

Introduction

The ready and enthusiastic agreement of so many of our Japanese colleagues to participate in an Anglo-Japanese geographical seminar is a most welcome development, and one that illustrates the long-standing fascination that Britain holds for Japanese geographers in general. This interest has led amongst other things to the appearance of many British geography publications in Japanese translation, and has been productive of substantial Japanese research on diverse British topics - witness, for example, the abundance of papers, in Japanese scholarly journals, on regional development planning in the U.K. and on aspects of industrial location in Britain.

Would that there existed a reciprocal interest in Japan among British geographers. For, and forgive me for beginning the first paper of the Seminar on a somewhat downbeat note, the fact is that ever since its late nineteenth-century acquisition of status as a university discipline, British geography has paid relatively little attention to Japan. Indeed it is not only Japan that has suffered neglect among British geographers but the whole of East Asia: over the last hundred years, well-informed British geographical studies of China have been depressingly sparse (Hong Kong, of course, is another question), while Korea - north and south - has been virtually ignored altogether. Later in this paper, I shall try to address the question of why Japan has been neglected, but to begin with, I should like to outline the progress that has been made since the end of the nineteenth century in British geographical research on Japan.

Background

In the years when university geography in Europe was in its infancy, it was not the British but continental scholars who first embarked on the serious study of Japan. In Germany, advanced studies of Japan appeared at a surprisingly early date: here I refer to the monumental works of J.J. Rein, professor of geography first at the University of Marburg and then at Bonn. Rein was an early authority on Japan and its economy. His writings (Rein, 1884 and 1889), weighty, thorough and immensely detailed, remain an invaluable source of information on Japan as she was at the dawn of her industrial revolution. French geographers of the late nineteenth and early twentieth centuries were perhaps less attracted towards Japan than their German colleagues, yet Japan came within the purview of Volume IX of *Géographie Universelle* (the volume entitled *Asie des Moussons*) where one may find an elegant and concise exposition on the country's geography, in both its physical and human dimensions (Sion, 1928).

If Germany and France provide us with examples of professional approaches, in Britain, by contrast, early discourse on the geography of Japan was of three main kinds: the 'capes-and-bays' descriptions of the country contained in elementary texts for school-room use; the accounts given by early scientific travellers to Japan that appear in the pages of the *Journal*, and *Proceedings of the Royal Geographical Society* (the forerunners of the present-day Geographical Journal) and of the *Transactions of the Asiatic Society of Japan*; and guidebooks written for well-heeled tourists. One suspects that the early school textbook descriptions of Japan generally did far more harm than good: the less satisfactory of these texts probably helped to cultivate, in the minds of generations of British school children, a lasting image of Japan as a land of geisha girls, rickshaws, and paper houses: the quaint and romantic Japan so beloved of Lafcadio Hearn and Pierre Loti. Whatever its provenance, this caricature image of Japan proved to be embarrassingly long-lived: its survival well into the 1950s prompted the late Ryujiro Ishida to write an excellent short textbook on the geography of Japan expressly designed to correct widely-held Western misconceptions (Ishida 1961).

Of the other two approaches to Japan, little need be said. Some of the early accounts - one is tempted to call them accounts of exploration - in JRGS and TASJ are potentially useful for a reconstruction of people-environment relationships in late nineteenth-century Japan (see, for example, Alcock, 1861 and 1862; Blakiston, 1872; Chamberlain, 1895), and the same is broadly true of Murray's *Handbook* (Chamberlain and Mason, 1903), by far the best of the early English-language guidebooks. Basil Hall Chamberlain, one of the joint compilers of Murray's *Handbook*, and an authority on Japanese literature and philology, was a Fellow of the Royal Geographical Society, as was T. Philip Terry, the author of the American equivalent of Murray's Guide.

With few exceptions, school texts, travel writing and guidebooks formed the sum total of British geographical writing on Japan until well into the twentieth century. The first substantial and analytical account of the geography of Japan to appear in the English language - J.E. Orchard's *Japan's Economic Position* (1928) - was by an American, and, indeed, Americans were to set the pace in Western geographical writing on Japan until well into the 1960s. This is not the place to comment at length on American contributions to the literature: suffice it to say that the chief milestones along the way were various papers by R.B. Hall Snr. (see, e.g., R.B. Hall 1931, 1937), a monograph by Smith and Good (1943), E.A. Ackerman's book on Japan's resource endowment (1953), and, of course, the magisterial survey undertaken by G.T. Trewartha, whose textbook, sadly dated though it might be, remains easily the most impressive treatment of the geography of Japan ever undertaken by a Western scholar. Trewartha's book (1945), surely a classic example of the genre of regional geography writing, was based on field work carried out in the 1920s and 1930s and on extensive collaboration with Japanese geographers. A much-revised and up-dated version, somewhat less comprehensive in scope than the first, appeared in 1965.

American geographers continued to make important contributions to the literature on Japan in the 1950s and 1960s (some of them, such as J.D. Eyre and D. Kornhauser belonged to an extraordinarily gifted cohort of American Japanologists who undertook their Ph.D research at the Okayama field studies centre of the University of Michigan). But while Americans were making useful and in some cases pioneering contributions, British geographers appeared to show little interest in matters Japanese. It is perhaps significant in this regard that the group of scholars trained in the Japanese language during the war years - a generation that was to be of formative influence in the postwar evolution of Japanese studies in Britain - did not include a geographer; more important still, perhaps, was the prohibitively high cost of travel (high in relation to the slender salaries of postwar geographers), a factor that discouraged frequent and lengthy visits to Japan.

Developments in the 1950s and 1960s

Nevertheless, strenuous efforts were made during the 1950s and 1960s to build bridges between the British geographical establishment and its Japanese counterpart. L.D. Stamp (LSE) and E.M.J. Campbell (Birkbeck College) represented the U.K. at the 1957 Regional Conference of the IGU in Tokyo - both became enthusiastic proponents of the need to forge stronger links with Japanese colleagues. In the late 1950s and early 1960s, M.J. Wise (LSE) and C.A. Fisher (Sheffield) visited Japan, met leading Japanese geographers, and established the basis for a long-lasting and fruitful connection with the world of professional geography in Japan. Two senior Japanese geographers - T. Kawashima and K. Murata - were subsequently granted long-term academic hospitality by the LSE, and the LSE-Japan link has been successfully maintained to the present day.

Charles Fisher, who is perhaps best known for his work on Southeast Asia, came into close - and sometimes painful - contact with the Japanese while serving as a prisoner-of-war in Singapore and in camps along the Burma Railway (Fisher, 1979). After the war, and despite the unpropitious nature of his first direct contact with things Japanese, he developed a strong scholarly interest in Japan, an interest that bore fruit in a number of academic papers and chapters in books concerned with Japan's position in the political geography of Asia (Fisher, 1950a, 1950b, 1950c, 1968, 1971).

Throughout his postwar career, Fisher worked hard to explain Japan to British audiences (often by way of lectures given with inimitable panache and bravura), and devoted much of his time and energy to cultivating links with Japanese geographers. He was the first Director of the Centre of Japanese Studies at Sheffield University, but left that post after a short while to become, in 1964, the first professor of geography at the School of Oriental and African Studies (SOAS). At SOAS, Fisher built up a department distinguished for its unique commitment to the geography of Asia and Africa. Within this department, two posts were created with specific reference to Japan: my own (I joined the staff in 1966) and that occupied by my colleague Richard Wiltshire (who came to us in 1979).

Paradigm shifts

By the mid-1960s, British geography was better placed than ever before to embark on the study of Japan. Useful contacts had been made with Japanese geographers, and senior and influential members of the profession were actively striving to promote a greater interest in Japanese geography. Moreover area studies were in vogue. British universities, with government support, began to show enthusiasm for the promotion of multidisciplinary research on areas outside the West, and new centres for this purpose were set up in Sheffield (Japan), Leeds (China), Hull (Southeast Asia), and elsewhere. And, of course, British public interest in Japanese affairs was growing quickly, for Japan in the 1960s was in the midst of its 'era of high-speed growth' (to borrow the phrase used in Japan): here was a country undergoing spectacular industrial transformation of a kind that put Britain's postwar economic performance in the shade. In the academic world, issues such as the nature and causes of Japan's postwar economic growth, the social and political contexts of Japanese industrialization, and the relevance of Japan's growth experience for less-developed countries began to attract keen attention.

Any innocent layman would be forgiven for supposing that geography was in the forefront of this growing interest in Japan: it might well be surmised that geographers, with their concern for explaining and interpreting the world in which we live, would be inexorably drawn into the study of contemporary Japan and, given the breadth of their interests, would preside over multidisciplinary work on the country. Alas, such was not the case. In geography, the opportunities for the study of Japan afforded by trends in the 1960s were generally squandered, and the number of British geographers fully committed to research on Japan remained disappointingly small. This is vividly illustrated by the membership of our professional area studies association - the British Association for Japanese Studies, formed in 1973. At the last count, in 1986, there were only four geographers in a total membership of 128. Another indication of interest in the field is the number of successfully completed Ph.D theses (completed, that is, by British nationals) on aspects of the geography of Japan. As regards the whole of the period since 1965, I know of only three examples nationwide - one at SOAS, one at Sheffield, and one at Manchester.

What went wrong? Part of the answer lies in the revolution that occurred within the discipline during the 1960s. The 'New Geography', heralded by the publication in 1965 of P. Haggett's *Locational Analysis in Human Geography*, constituted a major paradigm shift within the discipline, and whatever its manifest benefits in other directions, this shift was, I suspect, largely detrimental to the further development of the area studies approach. For many years following the appearance of Haggett's book, geographers became chiefly concerned not with demonstrating the uniqueness of place, but with the building and refining of geographical theory, using models that attempted to be as predictive and precise as those of the natural sciences (for some penetrating observations on the implications of this for area studies in geography, see Farmer, 1973). Out went the traditional 'regional geographies' whether at local or at national scales (and, in some cases, one must admit, it was a case of good riddance) and in came the quantitative revolution, model building, and systems analysis.

A further trend, and one that gathered pace in British geography in the 1970s, was the growth of interest in the relevance to geographical study of Marxist thought. This development brought us D. Harvey's *Social Justice in the City* (1973), D. Massey's *Spatial Divisions of Labour* (1984), and many other writings that sought to provide 'an understanding of space under capitalism' (Smith, 1981). It also helped to generate a new enthusiasm, among British geographers, for the study of Third World problems: the current strength of development studies within British geography owes much to the growth of socialist-inspired concern over the inequalities between the rich and poor countries of the world. Be that as it may, the burgeoning of geographical research that is informed by heightened political awareness has not worked to the benefit of Japanese studies - the 'capitalism' that has of late preoccupied many geographers in Britain often turns out to be either American or European capitalism, or indeed a capitalism of the mind: one that has no national peculiarities at all. Japan, with its very distinctive and, dare one say it, largely successful variant of capitalism, again seems to have been overlooked.

Constraints on the expansion of Japanese area studies

Ideological considerations apart, several other explanations can be advanced for the neglect of Japan among British geographers. One of these, clearly, is the language barrier. The fact is that apart from topics that fall wholly within the subject area of physical geography, no worthwhile progress can be achieved in research on the geography of Japan without recourse to the Japanese language. There exists a vast wealth of geographical literature in Japanese (there are, for example, six main disciplinary journals that carry articles only in Japanese, and one of these appears on a monthly basis). Then there are the numerous and substantial government publications, many of which, such as the annual white papers on land use and on environmental problems, are of direct relevance to geographical research. For work to be critical, well-informed and original, the Japanese disciplinary literature, and some of the literature in cognate fields, needs to be tackled and some of it mastered. And of course, it is not just language fluency that is a prerequisite for successful research: the investigation of most topics requires a working familiarity with the complex social political and economic realities that so often form the context of geographical enquiry.

All this takes not just an extraordinary degree of commitment but a great deal of time. Although the Japanese language is not an invention of the devil (as was fervently believed by many of the early Western missionaries in Japan), starting from scratch it takes a minimum of two years of intensive language work to acquire sufficient expertise to begin to make sense of a Japanese newspaper - and during that period, the aspiring student must somehow find time to read in the voluminous, and generally good quality English-language literature in cognate disciplines such as history, sociology, political science, and economics. In America, where a relaxed view is taken of the time needed to complete a Ph.D thesis, postgraduate students can meet all of these demands and spend lengthy periods in the field besides. But in Britain, most grant-giving bodies enforce strict limits to Ph.D completion periods - the current ideal, it seems, is three years for a Ph.D, and there are signs that before long, a two-year period might be insisted upon. The minimum amount of time needed to complete a worthwhile Ph.D on any aspect of the geography of Japan, is, in my view, four years.

And of course even if postgraduate students persevere and overcome all of these various barriers, an academic career in Japanese studies is nowhere near as remunerative as employment in Japan-related business and commerce. As Japan's economic presence in the U.K. grows, so British firms compete to attract intelligent graduates - and postgraduates - who are fluent in the Japanese language: in the City of London, for example, firms have been known to offer such individuals starting salaries that are in far in excess of the incomes of young university teachers. This consideration, more than anything else, explains why it is difficult to recruit good language graduates into postgraduate work in geography.

Recent trends

So far, I have painted a bleak and dismal picture of the standing of Japanese studies within British geography, and of the contributions that British geography has made to Japanese area studies. But it would be wrong to suggest that all is doom and gloom. Despite the constraints that I have mentioned, some useful work is being accomplished and there are signs, albeit faint ones, of a new growth of interest in Japan among British geographers.

On the research and publications front, the last few years have seen the appearance in Britain of two books and of numerous articles on aspects of the geography of Japan. Nothing comparable to Trewartha's regional geography has appeared, and I, for one, am a little dubious of the value of such textbooks: as D. Stoddart (1987) has pointed out, regional accounts are all very well but do not necessarily guide us in useful research directions. But regional geographies certainly have considerable didactic value, and in this regard it is pleasing to record that a well-informed and long-overdue book has recently been written primarily for use in schools (Macdonald, 1985). At a more advanced level, M. Hebbert (LSE) has collaborated with a Japanese scholar to produce a monograph on land development and planning in the Tokyo region (Hebbert and Nakai, 1988). This book, amongst other things, demonstrates the value of close cooperation between British and Japanese colleagues.

Meanwhile, papers have appeared on industrial location in Japan (Sargent 1980, 1987), on the geographical mobility of labour within Japan (Wiltshire 1989, 1990), and on Japanese direct investment overseas (Dicken 1987a, 1987b; Sargent 1990). At SOAS, we have recently begun research on geographical aspects of industrial restructuring in Japan, with particular reference to the iron and steel industry. A short article on this work has already appeared (Sargent and Wiltshire 1988), and further publications are planned - on the effects of restructuring on labour mobility within the steel industry (Wiltshire) and on local consequences of restructuring in the city of Muroran, Hokkaido (Sargent). In the longer term we hope to begin research on various aspects the geography of Hokkaido, perhaps in collaboration with scholars on the staff of Hokkaido University of Education, an institution with which the School already enjoys close links. Meanwhile a postgraduate student of the SOAS department is investigating aspects of the urban geography of Tokyo: his research, while not yet complete, has already yielded some useful preliminary conclusions (Waley 1990).

Not only is some progress being made in research on the geography of Japan, but the climate for such research has improved somewhat in the last few years. The extraordinary rise in Japanese direct investment in Britain (the U.K. has taken the lion's share of such investment within the EC) has brought Japan to the doorsteps of British geographers: those who have hitherto preferred geographically parochial parameters to research (including those who believe that geography, like charity, should begin at home) now find that the Japanese have become their co-parishioners.

Whatever may be the case among university and polytechnic colleagues, British school teachers of geography have suddenly been obliged to interest themselves in Japan. This is partly because of the increased Japanese presence within the U.K. and partly in response to syllabus changes. The British government is now moving towards the introduction of a national curriculum (the lateness of this initiative might occasion mild surprise in Japan, where national curriculums have been operation ever since 1872), and under the Education Reform Act of 1988, the proposed curriculum will include geography as a core subject. The geography syllabus, moreover, includes Japan as an area of study.

One hopes that the inclusion of Japan in the national curriculum for geography will lead, in due course, to a growth in the demand for university-level courses on Japan, and hence to an increase in the number of postgraduate students willing to tackle the challenges and opportunities afforded by the study of Japan. In the meantime, the Anglo-Japanese Geographical Seminar, involving as it does the exchange of ideas and information between professional geographers in both countries, will have an important part to play in raising awareness among British colleagues of the vast potential that Japan holds for geographical research. That there is much to be done on this score is beyond question.

References

Ackerman, E.A., 1953, *Japan's Natural Resources and their relation to Japan's Economic Future*, University of Chicago Press.

Alcock, R. (Sir Rutherford Alcock), 1861, 'Narrative of a journey in the interior of Japan, ascent of Fusiyama and visit to the hot sulphur-baths of Atami, in 1860', *Journal of the Royal Geographical Society*, 31, pp.321-356.

Alcock, R. (Sir Rutherford Alcock), 1862, 'Narrative of a journey through the interior of Japan', *Journal of the Royal Geographical Society*, 32, pp.280-293.

Blakiston, T., 1872, 'A journey in Yezo', *Journal of the Royal Geographical Society*, 42, pp.77-142.

Chamberlain B.H., 1895, 'The Luchu Islands and their inhabitants', *Geographical Journal*, Vol. 4, Part 5, pp.289-319.

Chamberlain, B.H. and Mason, W.B., 1894, *A Handbook for Travellers in Japan, including the Whole Empire from Yezo to Formosa*, London: John Murray. (Fourth edn.)

Dicken, P., 1987, 'Japanese penetration of the European automobile industry: the arrival of Nissan in the U.K.', *Tijdschrift voor Economische en Social Geografie*, 78, pp.59-72.

Dicken, P., 1988, 'The changing geography of Japanese foreign direct investment in manufacturing industry: a global perspective', *Environment and Planning (A)*, 20, pp.633-653.

Farmer, B.H., 1973, 'Geography, Area Studies and the study of area', *Transactions of the Institute of British Geographers*, 60, pp.1-14.

Fisher, C.A., 1950a, 'The expansion of Japan: a study in oriental geopolitics. Part I: continental and maritime components in Japanese expansion', *Geographical Journal*, 115, pp.1-19.

Fisher, C.A., 1950b, 'The expansion of Japan: a study in oriental geopolitics. Part II: The Greater East Asia Co-Prosperity Sphere', *Geographical Journal*, 115, pp.179-193.

Fisher, C.A., 1950c, 'Japan', in East, W.G. and Spate, O.H.K. (eds) *The Changing Map of Asia*, London: Methuen. (First edn) pp.321-331.

Fisher, C.A., 1968, '"The Britain of the East"? A study in the geography of imitation', *Modern Asian Studies*, 2, pp.343-376.

Fisher, C.A., 1971, 'The maritime fringe of East Asia', in East, W.G., Spate, O.H.K., and Fisher, C.A. (eds), *The Changing Map of Asia*, London: Methuen. (Fifth, revised edn)

Fisher, C.A., 1979, *Three Times a Guest*, London: Cassell.

Haggett, P., 1965, *Locational Analysis in Human Geography*, London: Edward Arnold.

Hall, R.B., 1931, 'Some rural settlement formations in Japan', *Geographical Review*, 21, pp.93-123.

Hall, R.B., 1934 and 1935, 'Agricultural regions of Asia. Part VII: the Japanese Empire', *Economic Geography*, 10, pp.323-347 and 11, pp.33-52, 130-147.

Hall, R.B., 'Tokaido, road and region', *Geographical Review*, 27, pp.353-377.

Harvey, D., 1973, *Social Justice and the City*, London: Edward Arnold.

Hebbert, M., and Nakai, N., 1988, *How Tokyo Grows: Land Development on the Metropolitan Fringe*, London School of Economics: STICERD.

Ishida, R., 1961, *Geography of Japan*, Tokyo: Kokusai Bunka Shinkokai.

Macdonald, D., 1985, *A Geography of Japan*, Woodchurch: Paul Norbury Publications.

Massey, D., 1984, *Spatial Divisions of Labour: Social Structures and the Geography of Production*, London: Macmillan.

Orchard, J.E., 1928, *Japan's Economic Position: The Progress of Industrialization*, New York: McGraw-Hill.

Rein, J.J., 1884, *Japan: Travels and Researches*, London: Hodder and Stoughton.

Rein, J.J., 1889, *The Industries of Japan, together with an Account of its Agriculture, Forestry, Arts, and Commerce*, London: Hodder and Stoughton.

Sargent, J., 1980, 'Industrial location in Japan since 1945', *GeoJournal*, 4, 3, pp.205-214.

Sargent, J., 1987, 'Industrial location in Japan with special reference to the semiconductor industry', *Geographical Journal*, 153, 1, 72-85.

Sargent, J., 1990, 'Japan's international economic strategy and its implications for Britain and the European Community', *Japan Digest*, 1, 1, pp.32-40.

Sargent, J., and Wiltshire, R., 1988, 'Kamaishi: a Japanese steel town in crisis', *Geography*, 73, 4, pp.354-357.

Sion, P., 1928, 'Le Japon' in Vidal de la Blache, P., and Gallois, L. (eds), *Géographie Universelle, Tome IX (Asie des Moussons, Premier Partie)*, Paris: Librairie Armand Colin, pp.189-237.

Smith, G.H., and Good, D., 1943, *Japan: A Geographical View*, New York: American Geographical Society.

Smith, N., 1981, 'Degeneracy in theory and practice: spatial interactionism and radical eclecticism', *Progress in Human Geography*, 5, pp.111-118.

Stoddart, D., 1987, 'To claim the high ground: geography for the end of the century', *Transactions of the Institute of British Geographers (New Series)*, 12, pp.327-336.

Terry, T.P., 1914, *Terry's Japanese Empire, including Korea and Formosa*, London: Constable.

Trewartha, G.T., 1945, *Japan: A Physical, Cultural, and Regional Geography*, Madison: University of Wisconsin Press.

Trewartha, G.T., 1965, *Japan: A Geography*, London: Methuen.

Waley, P., 1990, 'Tokyo: urban change in the Meiji and Taisho eras', *Japan Foundation Newsletter*, 18, 3, pp.16-20.

Wiltshire, R., 1989, 'Going around in circles: migration and internal labour markets in Japan', *Japan Education Journal*, 40, pp.9-10.

Wiltshire, R., 1990, 'Employee movement in large Japanese organizations', in Salt,J., and Johnson, J.H. (eds), *Labour Migration*, London: David Fulton.

Making Places, Making People

R.J. Johnston
University of Sheffield

The thesis that I want to develop in this paper is that we are what we are, as people, because of where we are and where we have been. This might sound like a case for environmental determinism - a geographical thesis that was discarded long ago. It is not such a case, however. What I will argue is that place, or locale, is a crucial environmental stimulant, not because of its physical characteristics, though they may have some indirect influence, but rather because of its human characteristics - the nature of the people who live and/or work there.

It might be argued that this is an entirely unexceptionable thesis, and therefore one that does not justify academic treatment: it is no more than common sense. What I hope to do, however, is not only sustain the thesis but also indicate its centrality to the understanding of many aspects of society, an understanding which is crucial to the development of society (in the proper sense of that term and not to be equated with some material indicator such as GNP per person). By illustrating the validity of the thesis and the case for its importance, I hope therefore to indicate some of the reasons why the study of geography is a necessary element in general education - which is why it is so important that we have gained it a place in the core curriculum being written for all British schools, and that we must not allow a narrow, instrumentalist, treatment of geography in that syllabus.

Place equals Culture

Basic to the whole of the thesis is Marx's much-quoted statement that 'men make their own history, but not in conditions of their own choosing' (here, perhaps, I should apologise for Marx's sexism). In other words, in terms of the classic 'nature versus nurture' debate over the formation of human characteristics, I come down on the side of 'nurture' - which is not surprising, since it would be extremely difficult to work in the educational profession if I did not believe that people's characteristics are made, not given. According to this view, people are created with mental equipment, with the ability to process information, but not with information itself (genetically encoded) nor with certain predispositions towards information. We live in a world of stimuli: how we respond to them - indeed, how we interpret them before responding - is not innate to us, according to the thesis, but is the product of our learning. And, of course, learning is a continuous process, one that involves us in constant contact with our environment - with the people in the various milieux that we occupy during our life-path.

Such a thesis runs counter to some of the arguments put forward recently by socio-biologists, who claim that the process of evolution can lead to certain behavioural traits being genetically transmitted inter-generationally. Thus whereas selfishness and altruism may not have been traits in early humans, Darwinian processes of natural selection may have led to a preponderance of people with such traits. It is beyond my scientific capability to assess the validity of such claims. What I would argue is that even if the general argument of the socio-biologists is accepted, the traits that are transmitted are extremely general, and how they are made operational is very much context-dependent. Thus selfishness may well be an inherited element of human nature, but how and when people act selfishly is a function of the milieu of the action and the history of the actor.

History is crucial to the argument, therefore. To many people, history and geography are separate disciplines, but only the briefest of considerations will indicate the futility of the divorce. History is about events at particular times and at particular places; geography is about particular places at particular times. Our histories - individually and collectively - are inextricably bound up with our geographies: what I am is a function of where I have been; what you will be is a function of where you were and are, which is why none of us will ever be the same even after this brief encounter, however limited its impact on you, if not me.

The importance of place in the early years of socialisation is clear, because the vast majority of people are raised in a spatially very restricted milieu among a very small number of people. We learn from those people, not just because of their didactic actions - important though those may be - but rather because they provide role models for us, patterns of behaviour that we adopt as our own. As we move out of those initial socialisation experiences, so both the spatial range and the social range of our contacts will widen. We see new role models, which we interpret in the context of our earlier appreciations and understandings. As a consequence, we may change the ways in which we respond to certain stimuli, so that as we follow our life-path so we change as people: this may be because we move to different places, and come into contact with different people, or it may be because different people come to our place - and we notice them.

Places change, continually, because the people in them change. At any one time, the nature of a place reflects the nature of the people there. Its culture - which is as good a term as any other to describe its nature - may be very homogeneous, because the great majority of the people there have been socialised in the same way and share a wide range of attitudes and beliefs. Alternatively, it may be very heterogeneous, comprising people from different backgrounds and providing conflicting role models for the socialisation process. Whichever it is, as people live in it, so they both reproduce and change it; as they re-create themselves, so they participate in the re-creation of the place. The result is a complex cultural mosaic, one which is forever changing and yet has within it a great deal of continuity. The culture of a place is not independent of its people - to believe that would be to believe in a form of cultural determinism - but it is something that they assimilate and sustain: to a great degree they can do no other, since individual and group survival is dependent on accepting the local cultural norms - in large part if not entirely.

Place and Power

So far, the picture I have painted is of a cultural mosaic into which people are socialised: all individuals play a part in the reproduction of that mosaic, but none control it, or seek to - it is simply the set of worlds they inhabit. But this is over-simplified, for clearly some people wish to exercise power over a culture, to promote certain ideas and to deny others: 'There is no alternative', to quote Mrs Thatcher. Why they wish to exercise such power at a fundamental level is a question I shall not address here (I am not concerned with whether it is genetically innate or socially learned). What I am interested in is that in wishing to exercise power in that way, what they are seeking to do is to change places - not in the way that David Lodge so superbly describes but rather to change the character of a place so as to promote individual and group goals.

The exercise of power is central to the operation of the societies that we know - both the capitalist societies that we live in and the socialist societies that are offered to us as alternatives. To the extent that the exercise of power involves the creation of milieux within which certain attitudes and beliefs (ideologies if you will) are fostered, so it involves creating places. The power that people want is power over people: they often achieve it indirectly by creating places with certain characteristics. Thus the exercise of power in society is an exercise in geography, an argument that I want to press here with three brief case studies, two taken from the United States and the other from the United Kingdom.

Three Case Studies

Sunbelt - Frostbelt, and the Cultural Geography of the United States

My first case study is a macro-scale one, and is concerned with major regional differences within the United States of America. One of the major features of that country's economic geography over recent years has been the shift of manufacturing activity from the old-established regions of the northeast (the area frequently termed the frostbelt, or snowbelt) to the south and west (the sunbelt). Many people have sought to account for this shift.

Why should the sunbelt have boomed in recent years, after many decades of stagnation on the periphery of the United States' economy? A whole host of reasons has been suggested. The southern states, for example, are considered more desirable for employers because wage levels are low and the workforce is relatively weakly unionised. They offer a more congenial climate than do the frostbelt states, too, and their expansion has been promoted by a great deal of government assistance, as, for example, in the growth of the defence and aerospace industries (Sale, 1975).

These and other reasons all go towards an account of sunbelt growth, but they don't penetrate the real reasons: why is the south weakly unionised? Why has so much federal money gone to the south? The answers to these questions must be sought in a deeper analysis of the south as a place, an analysis that involves an exploration of its culture and history. Such an analysis is a massive task: all I can provide here is a brief sketch.

According to Daniel Elazar (1985), the United States contains within it three major cultures, each of which has its heartland in a different region of the country. The first of these, the individualistic culture, is the one that we normally associate with the United States. The market place prevails, and the role of government is solely to sustain its operations; politics is a means to economic and social advancement for the enterprising individual. The moralistic culture emphasises the role of community and portrays government as a promoter of the good society; the role of politics is to temper individualism, by promoting moral concerns over those of the marketplace. Finally there is the *traditionalistic culture*, which is paternalist and elitist, distrusts government and democracy, and wishes to preserve a strictly hierarchical society.

Elazar has mapped the distribution of these cultures in some detail. Along the eastern seaboard, we see that each culture dominates a different segment of the country's heartland: the moralistic culture is characteristic of New England and, especially, the states of the Great Lakes basin which were settled by pioneer communities from northern Europe; the individualistic culture dominates the slice of the country inland from New York and Philadelphia, which was the heart of the country's first industrial revolution; and the traditional culture dominates the south, except in those states (notably Florida) which have received substantial numbers of retirement migrants (to the Costa Geriatrica) in recent decades.

It was in the heartland of the traditional culture, of course, that slavery was practised as part of the hierarchical, paternalist society. After the Civil War, when slavery was outlawed, a new set of social relations (especially race relations) was established by the defeated white supremacists, who instituted 'slavery without slaves'. This was done through the Democratic Party, which only whites could join and which so organised elections (particularly the registration of voters) that blacks were effectively disenfranchised. The result was one-party rule, and therefore one-race rule.

This one-party dominance became important in federal politics after 1930. In Washington, power in Congress has long been held by the senior Senators and Representatives, who are able to capture the important committee chairs for themselves. Seniority is a function of re-election, and since they were usually returned time after time, the Southern Congressmen came to dominate both the House and the Senate. After the New Deal was introduced and the federal government became much more active in financing the country's social and economic affairs, these powerful Southern Congressmen were able to channel massive amounts of federal money towards their constituencies, in the process now widely known as pork-barrel politics. They were able to promote local agriculture and then, from the Second World War onwards, the defence and aerospace industries.

The culture of the south therefore led to the development of a region which had both a quiescent, cowed labour force - with little democratic experience and appreciation of the benefits of unionisation - and a favourable political climate. Employers in the traditional manufacturing states were finding increasing problems with their workforces, who were both militant and costly, and the benefits of new transport systems (largely paid for by federal subsidies, notably through the Interstate Highway System) and technological change in production encouraged them to move into the region of traditional culture. That region, or place, had developed a particular set of social relations and political structures into which people were socialised. It remains one of the most conservative areas of the country - in its attitudes to unions, to equal rights (for women as well as for blacks) and so on. In a period of economic recession, that place became much more attractive to investors than places further north, where people were socialised into a more conflict-based approach to life. (Interestingly, some of the major investors in the sunbelt were the pension funds of the trade unions in the frostbelt, who in seeking the best returns for their capital were undermining the interests of their members!)

The growth of the sunbelt was a function of cultural variations between places, therefore. Today, the differential is being reduced somewhat, and the individualistic culture area is striking back. But the term has entered the language, and every country must have its sunbelt.

Exclusionary Zoning and the Creation of 'Tight little Islands'
My second case is also American, and refers to the organization of land uses in the suburbs of metropolitan areas. It clearly illustrates how people use space to manipulate the social milieu, thereby having a very strong influence on local patterns of socialisation, especially with regard to schooling.

Residential segregation is not peculiar to American urban areas: indeed, it is a characteristic shared by all cities in the capitalist world, which suggests that it is a response to a basic set of human impulses, even if the details of how the segregation is created and maintained vary from place to place. In brief, its origins lie in the processes of group identification that are found in complex societies. In situations where people live in close proximity to many thousands of others, and work, and travel, and conduct their daily lives in near-contact with many thousands more, most people create coping strategies that involve them identifying with one or more groups, within which they develop close contacts, and distancing themselves from others. Such strategies characterise many aspects of life, such as cliques and friendship groups at workplaces, in student halls of residence, and so forth.

Distancing can take a variety of forms but its most common involves spatial separation: like people cluster together and occupy separate areas from 'others'. Such spatial separation may involve the creation of fixed boundaries between the groups; it may involve informal boundary delimitation, as in the use of graffiti as markers of a gang's turf; or it may be entirely unstructured, and yet happen. My example here involves the creation of fixed, legal boundaries.

Why, in the context of residential patterns, should spatial separation be such a popular strategy? One reason is that modern society is too complex for us to cope with, certainly permanently, so that by identifying with a group which occupies a particular segment of the urban area one is defining a 'safe haven', a place where one feels comfortable and at home. What group? In most societies, the vague concept of class is the most frequent discriminating variable - defined on such criteria as occupation and income, from which other characteristics, such as attitudes and tastes, flow. Once such separation exists, people are then socialised within it, and the great majority accept it as a good way of ordering their lives. Few want to change it, which can reflect a combination of successful socialisation (or indoctrination) and an evaluation of the benefits of separation.

Occupational class is not the only criterion on which residential distancing is based. In many societies, ethnic divisions (many of which are linked to the occupational class structure) are also used - based on such criteria as race, language, and religion. People choose to live among people of the same ethnic background: or are socialised into believing that it is the correct choice. This aids their sense of identity with a particular group, and has a positive impact on their self-regard. It can also bring negative consequences, however, for separation leads to relative ignorance, because people from different backgrounds rarely come into close contact. Where contact is absent, behaviour is based on stereotypes which are part of the group culture. Such stereotypes are usually based on negative characteristics, as part of a process promoting the positive aspects of 'us' and the undesirable aspects of 'them'. For much of the time, the tensions implicit in these characteristics make for some unease: occasionally they result in the conflict that we have seen in so many cities in recent years.

The perceived benefits of residential separation are twofold:
1) It allows people to identify clearly with a particular group and to enjoy a common life style in an area which the group dominates, even if it doesn't have exclusive rights to it; and
2) It allows people to distance themselves from groups which they are socialised into thinking of not only as different but also as threats.
From these, two other benefits flow:
3) Because most (public) educational systems are based on schools drawing on localised catchments, then each school should be socially homogeneous and therefore provide an educational milieu which strongly supports that of the home and neighbourhood; and
4) Because housing is a privately-produced and privately-consumed commodity in most societies, the price of homes can reflect both the characteristics of the dwellings themselves and those of their milieux.

Such benefits particularly favour the affluent, those who are socialised to think that others are inferior to themselves and that contact with them will be harmful: such feelings are important with regard to such people's reactions to possibilities of 'mixed' schools.

Given that distancing is socially desired, how is it ensured? If societies were composed of groups that were mutually antipathetic, there would presumably be no problem. But they aren't, and some people in all societies aspire to be members of groups that they are excluded from. Further, in complex capitalist societies, social (and geographical) mobility is necessary and people are encouraged to aspire to higher status group membership - through educational and occupational 'success'.

Clearly a major way in which such residential separation can be promoted is through the operation of the property market, and this is certainly the case in American suburbs. But it isn't considered sufficient: property developers may try to increase their profits by introducing 'undesirables' to an area, and thereby alter its social milieu and the perceived quality of its schools, not to mention the property values of those already resident there. How can they ensure that doesn't happen?

A major vehicle for ensuring neighbourhood exclusivity, especially for the affluent, is the local government system. In most states, metropolitan local government is extremely fragmented: in 1977, for example, the Detroit SMSA was divided into 107 separate municipalities, of which only 15 had populations in excess of 50,000 and nine contained less than 1,000 residents - Chicago had 261 and St. Louis 194.

These municipalities are autonomous with regard to two important functions: raising taxes to pay for local services (excluding education in most cases) and controlling land use. Thus affluent people living in a suburban municipality pay nothing towards the costs of providing social and other services elsewhere in the metropolitan area. They avoid the subsidy-contribution that is supposedly the benefit of progressive taxes. They keep their areas exclusive by strict control of land uses. They can't have zoning regulations that say 'the poor shall not live here' - let alone 'the blacks cannot live here'. But they can make it too expensive for the poor (including middle income groups) to live there - and hence the blacks are excluded. This is done by density controls: by setting minimum lot sizes and maximum building sizes, and by excluding apartments.

By promoting exclusive residential zoning in this way, suburbanites also promote exclusive schools. Education is provided by separate bodies - school districts - and these too are many in number in the suburbs. In 1977, metropolitan Detroit had 108 of these independent authorities, which were separate taxing bodies: only one (the City) had more than 100,000 pupils and one had less than fifty.

In these ways, suburban Americans create closed communities for themselves, which those who are less privileged find it very difficult to 'invade'. Many court actions against such policies have been brought under the Fourteenth Amendment, which guarantees equality of treatment to all. Nearly all have failed, because the Supreme Court has ruled that discrimination on the basis of wealth is not unconstitutional and that although explicit racial discrimination is barred by the constitution, implicit discrimination is not. Some zoning laws go too far, as that in Cleveland which implicitly barred a woman from living in the same house as her two grandsons (who were cousins), but in general terms exclusionary zoning preserves the social separateness, school exclusivity, and property prices of the suburbs: local government boundaries are almost as effective as the Berlin Wall.

The Anatomy of a Strike

This final case study is taken from Great Britain, and refers to a very different situation. In 1984-85, the National Union of Mineworkers held a strike, not for higher wages, but in protest against a programme of pit closures being introduced by the National Coal Board: the focus of the strike, according to the Mineworkers' leader Arthur Scargill, was 'the protection of jobs and communities'. The strike was not solid, however. Some miners continued to work throughout and there was a slow drift back to work over the year-long dispute. More importantly, however, for the full period of the strike the majority of the miners in Nottinghamshire, Leicestershire and parts of Derbyshire remained at work and production was maintained at all of the pits in the area. Their refusal to strike was based on the lack of a national ballot of all members; local ballots there were against the strike.

Why should the miners in this part of the country stand against the NUM and its leader? A simple answer might be that the strike had little relevance to them. Their pits are among the most productive in the country and have very substantial reserves, so jobs in the area were not at risk. Why, then, should the Nottinghamshire miners strike to protect the jobs of others, thereby suffering hardship themselves (the union did not pay strikers anything), especially since a ballot had not been called and so miners as a whole had not been given a democratic say on the issue? In other words, a prosperous coalfield was opting out - just as the affluent suburbanites opt out from helping less prosperous others in American metropolitan areas.

But the explanation isn't quite that easy. Next door to the Nottinghamshire field is the equally productive and secure South Yorkshire field, where support for the strike (which started there, near Rotherham) was very strong: indeed any boundary dividing the two fields is arbitrary, and the NCB and the NUM have different lines separating Nottinghamshire from Yorkshire - with neither coinciding with the administrative county boundaries. So why did Yorkshire strike and Nottinghamshire not?

The answer can be found in the history of the two fields. The productive part of the Nottinghamshire field lies in the northeast of the county, in an area known as the Dukeries. It is on the edge of Sherwood Forest, and comprises a number of large landed estates owned by Dukes, hence the name. When these landowners discovered the fortune that lay beneath their land, they became mine-owners too: the way that they organized their miners was crucial to the development of the area's social and political structure (Waller, 1983).

If we look first at the way in which work was organized at their pits, we see that they used a method - the butty system - that was not unique to there (it was used in Kent too), but was very different from that in nearby South Yorkshire. The butties were subcontractors, who organised independent gangs of workers at allocated coalfaces, and paid those gangs according to their production. Thus the workforce at these new, large pits was divided up into a large number of separate, competing groups. In South Yorkshire, on the other hand, the workforce was organised as a single unit by the coal companies, and all were paid the same from the proceeds of total production.

These two different ways of organizing the social relations of production led to very different attitudes to unionisation. Unions were established in both areas, but that in South Yorkshire was much stronger and more militant because it represented a unified workforce whose individual and collective interests coincided. In the Dukeries, the union represented divided groups and the miners' individual and collective interests were not the same: hence miners there would not support a union which might take action that could threaten their individual interests. As a result, the local union was relatively mild in its attitudes and more accommodating to the wishes of the owners than was that further north. Both were affiliated to the National Federation, but during the General Strike of 1926 a breakaway Dukeries union (the Nottinghamshire Miners Industrial Union) was formed which was not affiliated (unlike the alternative Nottinghamshire Miners Association). It rapidly decided on a return to work and until 1944, led by George Spencer, operated a nonpolitical, largely conflict-free regime with the owners.

The particular circumstances of the social relations of the workplace in the Dukeries were paralleled by a further set of special factors in the sphere of reproduction - the home. The mineowners provided colliery villages for their workers - as was the case elsewhere. But these villages were carefully planned. So was the choice of residents, who were selected not only for their ability to work in the mines but also for their social fit: they were expected to participate in village sporting and religious activities, and were required, for example, to maintain their gardens to a high standard, and were not allowed to keep pets (with dismissal as the penalty). Most of the villages were isolated, too, and the miners' families had to shop at the stores run by the mineowners.

More so than in South Yorkshire, then, the miners of the Dukeries were dependent on their employers, and felt that militant stands against them were not productive. This was reflected in local politics as well as in union affairs. Most of Britain's coalfields are historic centres of the growth of the Labour Party, but the Dukeries is not. Indeed in local politics the Labour Party was notable by its absence; there was no mobilisation of the miners by the party, and the pit villages were represented on local and county council and in parliament by Conservatives.

The basis for all this changed with nationalisation in 1947, of course. The NMIU had by then joined the NUM; the coalowners were dispossessed; national working and wage agreements were slowly introduced and the Labour Party began to organise and represent. But the attitudes bred in those two decades remained, and the younger generations socialised to them were similarly conservative (with a small 'c') and remained relatively non-militant unionists. Hence the attitudes displayed in 1984-85, attitudes that were reflected later by their break-away from the NUM to form the Union of Democratic Miners, which British Coal recognises but the Trade Union Congress does not. And the parliamentary constituency for the Dukeries - Sherwood - returned a Conservative MP in both 1983 and 1987 while nearby Mansfield became a marginal Labour seat.

This case study once again shows the importance of appreciating local culture if one is to understand contemporary events. One must not be deterministic about it, of course, and the culture can be changed. South Wales was the strongest local union to support the strike. But then, in 1986-87, the local union negotiated a new six-day week with British Coal, which was prepared to open a new pit with over 600 jobs in return. Five-day working has long been a central policy plank of the NUM, which opposed the South Wales action. But faced with acceptance or no jobs, the South Wales miners voted for jobs, and thereby changed a major element of their local culture. At the same time, the UDM and the NUM were contesting the right to represent miners at a new pit to be opened near Coventry, where once again six-day working was being insisted on by British Coal.

Memories are long in such local cultures, and men who broke the 1926 strike are still today denied entry to the miners' clubs in South Yorkshire. Here the divisions created in 1926 and recreated in 1984 will remain deeply etched in the social geography of England. New divisions were created in 1984, too. Within the Yorkshire coalfield are two separate administrative counties with their own police forces. That in South Yorkshire adopted a very high profile approach and there was a great deal of pitgate conflict - characterised by riot shields and cavalry charges. The result is continued great bitterness and tension against the police, exacerbated by the tactics adopted in Nottinghamshire (where pits were regularly picketed by South Yorkshire miners) that were designed to stop incomers reaching the pitgates. In West Yorkshire, on the other hand, the police took a very low-key approach with little hostility and no use of riot gear; as a result there is much less of a legacy of bitterness in local community relations.

In Summary

The three case studies that I have presented here only skim the surface of the great range of examples that can be brought forward to illustrate and sustain the basic thesis that I have developed. People are made in places; they are only able to develop their particular human attributes because they are in contact with others who transmit to them not only information but also, and much more importantly, ways of interpreting and using information. Reality is a social construction, and most realities are constructed in place-bound societies.

If we accept the basic thesis, then a number of conclusions follow. First, we are given insights into a whole range of social processes that operate at many spatial scales. Some of these may involve what we might call the creation of 'informal cultures', but others are much more formal. Nationalism is a very good example of an element of local culture (at the scale of the nation-state) which is propagated by powerful groups within a society to promote their own goals - by legitimating them with the population at large.

A second conclusion, which in the present context draws particularly on the second of the case studies, is the realisation that space is not just a stage on which society enacts its plays; it is a resource that is manipulated within societies as part of the script. To achieve certain ends, people create places, and in certain situations that involves defining those places with boundaries and then being prepared to defend them. National boundaries are obvious examples of that process, but so are the six-foot fences that so many New Zealanders erect around their quarter-acre pavlova paradises.

The third conclusion follows from that, for if space is a crucial societal resource then clearly the discipline that focuses on space - geography - is an important one. One's attitude towards the discipline might be that of a final honours degree student whose examination script I read recently: he wrote:
I should point out that I don't believe that geography should exist, but for the purpose of this paper please assume that it does.
Whether geography exists or not, its subject matter is central to the understanding of society.

Finally, and most importantly, the thesis underlines what we already know - that the world we have made for ourselves and are making for others is an extremely complex place. To understand it, and to transmit that understanding, is crucial, because unless people understand it, they will not be able to control it and particularly their own role in it. We must all appreciate that the place is many places, that we made them, and that we are made in them.

References
Elazar, D.J., 1985, *American Politics: A View from the States.* New York: Harper and Row.
Sale, K., 1975, *Power Shift.* New York: Random House.
Waller, R.J., 1983, *The Dukeries Transformed.* Oxford: The Clarendon Press.

Counterurbanization and the urban-rural manufacturing shift

Paul White
University of Sheffield

Introduction

It has become accepted wisdom in recent years that two major new trends in regional development started to occur in the countries of the capitalist world sometime during the 1970s. Economic geographers have identified what they have called the 'urban-rural manufacturing shift' (Keeble, 1984; Watts, 1987), for which evidence was already accruing in the United Kingdom and the United States as early as the late 1960s (Keeble, 1968; Fothergill and Gudgin, 1982). By the mid-1980s the phenomenon had spread throughout much of Western Europe, with particularly striking patterns of change in regional fortunes occurring in France (Aydalot, 1986) and in Italy (Brusco, 1986): the significance of new firms in the growth of manufacturing in rural regions was especially great.

At the same time population geographers were becoming aware of major shifts in the regional patterns of population growth and, in particular, of the migration component of that growth. The phenomenon of a 'turnaround' in rural populations, with decades of slow population loss being suddenly replaced by in-migration and growth, was first recognised in the United States in the early 1970s and led Berry (1976) to coin the word 'counterurbanization' to denote the pattern. In much of Europe this new pattern of population change had to await the censuses of the years 1980-82 for full investigation, but since then an impressive body of empirical evidence has built up to affirm the generality of this process of population deconcentration from urban areas and 'regrowth' in many rural regions (Champion, 1989). Nor is counterurbanization confined to North America and Western Europe. Hugo (1989) has shown the existence of the phenomenon in Australia, whilst in Japan Tsuya and Kuroda (1989) have argued that the pace of population concentration in the major metropolitan centres is slowing down, although here they see little real chance of 'population redistribution beyond the metropolitan boundary ... unless conditions are favourable for the location of industry' (p. 228).

This last point, related to the Japanese case, brings together the two phenomena of urban-rural manufacturing shifts and counterurbanization, and suggests causality between the two. Because the two phenomena have co-existed temporally it is attractive to suggest that they are mutually interdependent. However, the evidence for such an assertion is by no means strong. In particular, there are a number of possible explanations for the individual phenomena which do not depend on the existence of the other major trend.

Evidence from Western Europe

Counterurbanization in the United Kingdom was first described by Champion in 1981. The pattern as it appeared then, and its evolution in the following years, consists of a clear decline in the populations of the great metropolitan centres such as London, the West Midlands, Merseyside, Greater Manchester, South and West Yorkshire, and Central Scotland. At the same time there has been population growth in many predominantly rural regions of long-standing decline such as Central Wales, the Anglo-Scottish borders, and Northern Scotland.

However, if instead of considering aggregate population changes only the economically active population is examined (Champion et. al., 1987) it is clear that many rural areas that have experienced overall population increases have also seen a reduction of their labour forces. Clearly the links of rural population change to rural employment change are not very direct.

If we turn elsewhere in Europe evidence can also be found of areas where economic change and population change are not in phase. In France, for example, analysis of the 1982 census results shows that throughout much of the northern part of the country, from the borders of Brittany across the Paris Basin to the Swiss frontier, the pace of population change has been more favourable than the pace of employment change, leading some commentators to suggest that what has been happening is an extension of long-distance commuting (Noin and Chauviré, 1987) rather than a real turnaround in regional economic patterns. The phenomenon of rural revival originating in the arrival of non-employed in-migrants has been confirmed as being of great significance for many regions of France, particularly in the rural west (White, 1990), whilst at the level of an individual region Winchester (1989) has demonstrated the tenuousness of the connections between the growth of new manufacturing industry and the regrowth of the rural population.

Together with evidence from other European countries, doubt can clearly be cast on explanations of urban-rural manufacturing shift and counterurbanization that involve the mutual interdependence of the two phenomena.

Towards explanations

Economic geographers' explanations of the manufacturing shift towards rural regions draw strongly upon the processes leading to the creation of a 'new spatial division of labour' (Massey, 1979), and in particular the attraction of restructured manufacturing industries to new, often rural, sites for reasons concerned with labour factors (Watts, 1987; Peet, 1983). If this is, in fact, the case, the likely implications for counterurbanization trends are somewhat equivocal. It might be expected that the move of manufacturing into rural regions should reduce the rate of labour loss from such areas through out-migration, but the creation of a net inflow of population is more problematic, and might depend upon the in-movement of skilled workers, managers and their dependants to the areas in which the new plants are located. In this respect the residential attractiveness of rural regions for managers and entrepreneurs has been suggested as a contributory factor in rural manufacturing relocations.

In contrast with the general consensus on the causes of the manufacturing shift, there has been much greater disagreement about possible explanations for counterurbanization. One of the first discussions of the competing theories was provided by Court (1984), and a more recent review, emphasising the French literature, is provided by Winchester (1989). In total it is possible to identify five major lines of explanation as follows:

Counterurbanization theory

The emphasis here is on a voluntarist explanation of the reverse migration flow out of large centres. This holds that population movement into rural areas is driven by the evaluation of residential attractions of different types of regions and localities. Such a process would therefore be people-led, with the possibility of housing and job market effects following on. However employment, or at least income considerations play a role in any migration process, with the result that the only two major groups in society who can pursue real choice of residential location at this regional scale are either the self-employed, or the retired. In fact the latter group certainly plays a very significant role in migration to many rural regions throughout the developed world, and a role which is likely to increase in the future with the progressive ageing of national populations. The areas of attraction for such movement, and for self-employed groups who are often 'dropping out' of the urban 'rat-race' and trading down occupations to work for themselves (Jones, Caird and Ford, 1984), tend to be peripheral rural or coastal regions, so that this explanation of the general counterurbanization trend may only be of relevance in certain locations.

Suburbanization theory

This would argue that what has been witnessed is not a new process of 'counterurbanization' but a continuation of the phenomenon of urban spread, albeit in a more diffused and dispersed form. This argument has some attraction in many Western European countries with high population densities such as Belgium, the Netherlands or Denmark (Burtenshaw and Court, 1986) where the identification of real urban-rural breaks is problematic. It has also been suggested (above) as operating in Northern France.

Neo-classical economic theory

This suggests that migration should occur according to the tenets of economic rationality. It has been strongly criticised by Fielding (1985) as being inappropriate to counterurbanization flows precisely because here net migration flows are to lower wage regions. However, neo-classical economic theory may have some value in two circumstances: firstly if the labour market is disaggregated into occupational and skill sectors where it may be that the economic benefits of a rural location for certain sectors do in fast outweigh the urban rewards; and, secondly, in the limited number of cases where resource exploitation creates a power of labour attraction to a previously impoverished region. An example of this latter case would be the population growth in many parts of the Scottish highlands and islands associated with the oil industry during the 1970s.

State intervention theory

This would ascribe population movement into rural regions as being a response to government regional policies. There is, however, an unexplained time-lag present in this argument, since in many countries such regional policies have been in place for many decades before any such effects came to be discerned. Areas where state regional policy has been a contributory factor have probably only been where it has operated alongside some other cause (Spooner, 1972).

New spatial division of labour, or neo-Marxist, theories

These are the theories already outlined as used by economic geographers to explain the relocation of manufacturing functions into new regions.

Discussion

It should be clear from the consideration of these five theories that there is no necessary mutual exclusivity about them. There is no logical reason why we should accept one and reject the others. Just as urbanization is a complex process with a wide variety of economic, social, cultural and political ramifications and causes, so counterurbanization should be seen as an equally complex phenomenon. Monocausal explanations are likely to be misleading and to oversimplify complex realities.

In particular, different explanations for the rural population turnaround are likely to be applicable in different regions. In some areas it is retirement migration, in line with 'counterurbanization theory', that is likely to be the predominant factor, and this can clearly be suggested as the explanation of the pattern is the United Kingdom, described earlier, where rural populations may be increasing in many regions but with continued decline in the size of the employed group. The 'new spatial division of labour' argument has strong backing for certain regions, for example in central Italy where the 'Emilia' model of regional economic growth has become well-known as a classic form of post-Fordist industrial organisation (Adamo, 1986). And the 'suburbanization theory' has, as already suggested, strong relevance in other regions, particularly in the hinterlands of the great metropolitan centres.

The urban-rural manufacturing shift and the phenomenon of counterurbanization have both played a vital role in changing the fortunes of individual regions in the capitalist world during the past twenty years. The co-existence of these two trends in time should not, however, lead us to too simple a view of their causal inter-relationships. Counterurbanization can occur without manufacturing shift, as the experience of many regions in Western Europe can testify. Manufacturing shift may be a sufficient cause for counterurbanization but it is not a necessary one. Deeper structural causes of an economic, political, social and demographic character underlie both of these two important recent trends. The most significant links probably occur through a general re-evaluation of cities - a re-evaluation carried out both by corporate entities and by individuals, in which the old urban centres are increasingly seen as losing their attractiveness for investment and residence.

References
Adamo, F., 1980, 'The urban crisis in Italy'. In Heinritz, G. and Lichtenberger, E. (eds.) *The Take-Off of Suburbia and the Crisis of the Central City,* pp.207-221. Wiesbaden: Steiner.
Aydalot, P., 1980, 'The location of new firm creation: the French case'. In Keeble, D. and Wever, E. (eds.) *New Firms and Regional Development in Europe*, pp. 105-123. London: Croom Helm.
Berry, B.J.L., 1976, 'The counterurbanization process: urban America since 1970'. In Berry, B.J.L. (ed.) *Urbanization and Counterurbanization*, pp. 17-30. Beverly Hills, California.:Sage.

Brusco, S., 1986, 'Small firms and industrial districts: the experience of Italy'. In Keeble, D. and Wever, E. (eds.) *New Firms and Regional Development in Europe*, pp. 184-202. London: Croom Helm.

Burtenshaw, D. and Court, Y., 1986, 'Suburbanization or counterurbanization: the case of Denmark'. In Heinritz, G. and Lichtenberger, E. (eds.) *The Take-Off of Suburbia and the Crisis of the Central City*, pp. 54-69. Wiesbaden: Steiner

Champion, A.G., 1981, *Counterurbanization and Rural Rejuvenation in Britain: an Evaluation of Population Trends since 1971*. Seminar Paper 38, Department of Geography, University of Newcastle-upon-Tyne.

Champion, A.G. et al., 1987, *Changing Places: Britain's Demographic Economic and Social Complexion*. London: Edward Arnold

Champion, A.G., 1989, (ed.) *Counterurbanization: the Changing Pace and Nature of Population Deconcentration*. London: Edward Arnold.

Court, Y., 1984, *Counterurbanization: A Review and Bibliography*. Department of Geography, Portsmouth Polytechnic.

Fielding, A., 1985, 'Migration and the new spatial division of labour'. In White, P.E. and van der Knaap, G. (eds.) *Contemporary Studies of Migration*, pp. 173-180. Norwich: Geo Books.

Fothergill, S. and Gudgin, G., 1982, *Unequal Growth: Urban and Regional Employment Change in the UK*. London: Heinemann.

Hugo, G.J., 1989, 'Australia: the spatial concentration of the turnaround'. In Champion, A.G. (ed.) *Counterurbanization: the Changing Pace and Nature of Population Deconcentration*, pp.62-82. London: Edward Arnold.

Jones, H., Caird, J. and Ford, N., 1984, 'A home in the Highlands', *Town and Country Planning*, Vol 53, pp.326-7.

Keeble, D., 1968, 'Industrial decentralization and the metropolis: the north-west London case', *Transactions. Institute of British Geographers*, No.44, pp.1-54.

Keeble, D., 1984, 'The urban-rural manufacturing shift', *Geography*, Vol 69, pp. 163-6.

Massey, D., 1979, 'In what sense a regional problem', *Regional Studies*, Vol 13, pp. 233-43.

Noin, D. and Chauviré, Y., 1987, *La Population de la France*. Paris: Masson.

Peet, R., 1983, 'Relations of production and the relocation of United States manufacturing industry since 1900', *Economic Geography*, Vol 59, pp. 112-43.

Spooner, D.J., 1972, 'Industrial movement and the rural periphery: the case of Devon and Cornwall', *Regional Studies*, Vol 6, pp. 197-21.

Tsuya, N.O. and Kuroda, T., 1989, 'Japan: the slowing of urbanization and metropolitan concentration'. In Champion, A.G. (ed.) *Counterurbanization: the*

Changing Pace and Nature of Population Deconcentration pp. 207-229. London: Edward Arnold.

Watts, H.D., 1987, *Industrial Geography*. London: Longman.

White, P.E., 1990, 'Labour migration and counterurbanization in France'. In Johnson, J. and Salt, J. (eds.) *Labour Migration*, pp. 99-111. London: David Fulton.

Winchester, H.P.M., 1989, 'The structure and impact of the postwar rural revival: Isère'. In Ogden, P.E. and White, P.R. (eds.) *Migrants in Modern France*, pp.142-159. London: Unwin Hyman.

The Challenge of Place Description

Douglas C.D. Pocock
University of Durham

Geography, literally defined as earth description, has always focused on the distinctiveness of the several parts that make up our world. Indeed, a passion for geography, no less than the urge to travel, often originates from the very distinctiveness of places. Such distinctiveness, perhaps readily recognisable, may yet be elusive to define, thus deceiving those who would deem the task to be 'mere' description. It is, in fact, no simple task, no superficial exercise, to wrestle with the appearance and nature of places. Reality is inevitably multifaceted and is filtered by people's varying sensitivity, experience, knowledge, motives, and so on. The exercise of place description is illustrated here with reference to the city of Durham.

At one level, the description of Durham may be accurate and precise. Thus, maps show the city to be one of eight districts in the county, situated centrally and extending some 20kms east to west by about 12kms north to south. A tightly-delimited Central Urban Area, astride the county's main north-south road and rail links, contains approximately half the district's population of 82,000, having more than doubled in its built-up area since 1945. Other pertinent statistics complete this type of record. Employment in services, for instance, at 75 per cent of the total, is considerably above the regional and national average and reflects its role as a county town, market centre, educational and cathedral city. The proportion in manufacturing is, correspondingly, little more than half the national average. Little wonder, therefore, that the proportion of households in the socio-economic groupings of professionals, managerial and non-manual is also well above the national average. Other indices would confirm Durham as a particular kind of settlement, which might be summarised as being in, but not of, the North-East, but such a statistical record, however accurate or complete, can only reveal so much.

Complementary to any statistical or tabular record is a description of the key physical components. Here, Durham's modest size might be the first feature - and a cause of surprise if beforehand one has used the cathedral as a touchstone of expected size. Boundedness is a related, second feature. The city is a distinct entity; it does not fray at the edges; it does not stutter into being through suburbia. The monumental climax is its most distinctive feature. Upstanding and outstanding, the cathedral is the tallest, grandest building on the highest site. Nature formed the peninsula, which rises 30 metres above the encircling river; man exploited the potential by raising a stone monument from the rocky pedestal by at least an equivalent height. It is a dramatic climax. In complete contrast, the townscape is domestic in scale, organic in layout and varied in age. It rises and falls, winds and staggers as if in sympathy with the river course and topography. Greenness is a final, all-pervading feature. The countryside, even the working countryside, approaches and appears to enter the very heart of the city. Important here is the broad wedge which accompanies the river on its entire course through the city.

A pictorial record - description made visible - may further the appreciation of place. Such illustrations are particularly welcome in the instance of Durham, where 'long views' have attracted artists, scholars and travellers over the centuries. In the 17th century Hegge considered the visual impact of the city and its setting could save the pilgrim a trip to Jerusalem; in the 19th century Ruskin considered the view of the city as one of the wonders of the world; in the present century Pevsner ranked the views as one of the architectural experiences of Europe. Any possessed of a camera today are tempted to make their own effortless record, although replication of postcard scenes in conditions of high summer run the risk of an incomplete grasp of the subject. Inclusion of photographs taken at dawn, dusk, night-time, during inclement weather or at different seasons begin to 'fill out' the character of place by showing it not to be a static phenomenon, but a living entity with varying moods.

Atmosphere and emotion which adhere to place may be uniquely unlocked through sound - natural, man-made and machine-generated. Sounds enrich the experience and contribute to the 'thickness' of place. They are evocative and suggestive, thereby heightening anticipation or triggering memory. A sound portrait therefore has unique powers of depiction, which are able to 'move' any hearer. In Durham the sonic focus reinforces the visual climax. The cathedral clock marks the passage of time, as do bells for the daily services and evening curfew; triumphant pealing marks Sundays and major festivals. The surrounding river and steeply-wooded banks echo to its own tunes - notably of birdsong and boating, again with crescendoes at regatta time and music-on-the-river. Bands and festivities associated with the Miners' Gala, events in the Market Place and the Babel of foreign tongues of visiting tourists are other prominent strands which might be woven together to form a sound portrait of place (Pocock, 1987).

Place, then, is more than the physical or visual, more than can be mapped, tabulated, illustrated or described. It is always more than the sum of its component parts. Thus the quality of Durhamness springs not only from the heightened degree to which

142

particular features are present but from the pervasive co-existence of opposing features. A series of dualities, holding in tension the two opposites, can be recognised. Thus, Durham is spiritual *and* secular. At its heart are cathedral *and* castle. Even the cathedral itself, in Sir Walter Scott's well-known words, is 'Half church of God, half castle 'gainst the Scot'. Again, Durham is nature and culture. It is town *and* country: a city in a garden: a large temple in the middle of greenery. Durham is monumentalism *and* the domestic. It speaks of master masons with international ideas *and* of more modest craftsmen and the vernacular. The city speaks of one time period, yet is *timeless* in what it proclaims. As the capital of the county, it belongs to the people of the county, who have considerable pride in their town, in being Cuthbert's people. Yet, Durham belongs to *British* history, *western* civilisation, the *whole* of Christendom.

Durham as a place is above all a focus of meaning and experience in which its citizens and all who come into contact variously participate. Above the level of personal biography and private worlds, the city stands as a physical manifestation of a significant event in our collective history - confirmation by the conquering Normans of a site chosen for, or by, the North's most famous saint, a holy man from Holy Island. It is a physical manifestation of a story, a story with a quality of legend concerning St Cuthbert. Moreover, story and history have been certified for us in writings, verse and paintings. Knowledge of the past presence of famous men increases the 'thickness' of place; today we are able to walk where they walked, stand where they lingered, see what they saw: *our* experience is certified thereby. (No landscape, it may be held, is truly rich until it has been certified in this manner. It is particularly so when the story has the quality of legend or myth). In consequence, Durham is suffused with a spirit of place which, perhaps, ultimately only the brushwork of a Turner or the music of a Tavernor can attempt to convey.

Reference

Pocock, Douglas, 1987, *A sound portrait of a cathedral city*. Department of Geography, University of Durham.

Whisky, Landscape, History

I. G. Simmons
University of Durham

On a visit to Hokkaido in 1972 I was shown the country around Otaru. My guide telephoned his home to see if his wife was in, for she was a noted exponent of *cha-no-yu*.. Alas, she was out. No matter, said my guide, we will have a whisky ceremony instead: I know the manager of the Nikka distillery. The distillery was (and I hope still is) a replica of a Scottish distillery: built in granite in the style of a 19th century castle of that country. I was told that in its early days even the peaty water used for the whisky was imported from Scotland.

If we start in Scotland and Hokkaido, then there is no need to finish with that comparison, for others can be made. Both the UK and Japan consist of sets of islands off the same major continent (1); both are in the temperate latitudes of the globe (though the UK is in general rather cooler than Japan) and at one time had large areas of both deciduous and coniferous forest; both had a long period of development of an agricultural economy under feudalism which was largely directed towards local self-sufficiency for food, and both have experienced industrialisation since the 19th century. Have we then anything to learn from both our similarities and our differences in the consideration of how humans and the non-human world interact, both yesterday and tomorrow?

The hermeneutics of landscape and history
Looking back at the landscapes of Britain and of Japan (in both their rural and urban forms) and at the environmental history of the two nations, can we draw any meaning from these sights and that knowledge? As scholars, can we interpret these phenomena in a way not accessible to the tourist or to those who are blind to the nuances of their surroundings or who cannot read a landscape as a textual expression of the structures of a society as well as its past? We certainly can, but not necessarily in one way only. Two obvious alternatives present themselves immediately.

The scientific world-view
Much has been written, and no doubt will continue to be written, about the world-view (often labelled by the German word *Weltanschauung*) which has, since the 18th century, come to be characteristic of the West and by extension of most parts of the world where any modernisation of the economy has taken place. Here it is sufficient

to say the viewpoint involves assent to the ideas that nature is a set of resources for human use, that progress is linear and will go on for ever, that growth of most kinds but especially of economic indicators like production, consumption and GNP is desirable and indeed should be sought at more or less any cost, and perhaps of most interest for our present purpose, that the human species is qualitatively different from the rest of the living matter of the planet, to say nothing of the non-living materials.

The major intellectual outgrowth of this world-view is without doubt that of the floresence of science. This depends epistemologically on acceptance of the notion that the observer (always a human) can be separate from what is being observed and that any observations and measurements that are made are (i) real, i.e. that they relate to phenomena which are actually present in the world and not simply in the human mind; and (ii) that they are totally objective and thus free from any selectivity introduced by human values or cultural biases. The dominant mode of investigation has been that of reductionism, which is the breaking-down of complex systems into simpler ones that are more amenable to measurement and investigation. Thus an organism is studied in terms of its organs, the organs in terms of their tissues, and the tissues in terms of their component cells. The cells are collections of molecules, and even these may be better understood if their atomic characteristics are known. This whole paradigm has been immensely successful as a way of making temporal predictions about the behaviour of both living and non-living systems. The space programmes of the super-powers and the contraceptive pill both owe their success to this methodology. The branch of science most concerned with the relations of humans with their surroundings is not always easy to identify. The most obvious candidate is Geography but since the 1950s many of its practitioners have been more concerned with making it more like the other social sciences with which it shares a phenomenology (e.g., sociology, economics, political science) though the introduction of scientific method has been a priority. Anthropology has had its excursions into cultural ecology but they do not seem to have gained the centre of the stage. Economics likewise has its branches much concerned with the economics of resources and the environment but its poor record of producing temporally accurate predictions has diminished its claim to be a science.

Even ecology, like most other branches of science and social science, can be seen as having one long-term aim: that of control (2). It has become one of the ways in which western and western-influenced societies have sought to dominate their natural environment, largely in order to gain resources from it and to re-make it in this image. A frequently used term for the use of knowledge in this way is *technocratic*; it refers to the underlying world-view (*v.s.*) that nature is a set of materials for human use.

On the other hand, the very fact that ecology is interested in wholes (of which the Gaia hypothesis is currently the most popular) links it to the philosophical concept of holism and thence of course to very non-scientific views indeed, for we are in the area when the objectively verifiable knowledge produced by science is set alongside intuitive knowledge produced by an individual person. It is not surprising, then, that ecology has come to mean different things in the world: to mean an attitude to nature which is protective and non-exploitative and even to be the name of a political party at one time in the UK, though the 'Ecology Party' has now renamed itself the 'Green Party'. So ecology will indeed provide a framework which will enfold both man and nature and their interactions but many have found it in its pure form, *qua* science, to be an incomplete guide to the perplexities of living on Earth in the 20th century. We ought now to turn to some of the other choices on offer.

Western phenomenology

A recent book by N. Evernden (3) gives some idea of the potential of the idea in the fields of interest of this essay. He takes us, among other things, into an account of the relevance of the German philosopher Martin Heidegger (1889-1976), a pupil of Husserl, whose writings are very influential though extremely difficult to understand. Following Husserl, he advocates going beyond the categorisation of the world that is so characteristic of science since Descartes and, as Evernden puts it, 'returning to the things themselves, to a world that precedes knowledge and yet is basic to it, as countryside is to geography and blossoms to botany.' What is important is experience: we shift focus from the reality of the world in science's terms to the meaning of the world and, further, a world in the division into subject and object is trivial, for they are always enfolded in one another. Heidegger was convinced that humanity had set off on the wrong road in ancient Greece when the question 'what does it mean *to be*?' ceased to be important. The question for Heidegger is to describe what Being means to humans and the answer (in part at least) is that it means relating to the rest of the planet through 'care'. The boundary of the self is not the epidermis of the body but a gradient of involvement with the world, analogically considered as being like a magnetic force-field. So, Being (in German, *Dasein*) represents not just a person but a field of care, for a Being is a being-in-the-world where the world is what makes each person human because it is can be understood by them.

146

In his later years, Heidegger was much inclined to a poetic view of language, and no Japanese reader will need to be told of the vital cultural role of the artist (whether with words, brushes or music) in showing us the importance of the way in which we perceive (or perhaps recognise) our surroundings. We might merely pause at this point to mention that such traditions are long-standing in the UK as well. To an Englishman, the music of Edward Elgar or Ralph Vaughan Williams, for example, can conjure up visions of the rural landscape of field, hill and small woodland as vividly as word-pictures or photographs. Poets and novelists of landscape and place are well-known and receive critical attention for this aspect of their work. Modern writers like the poet Ted Hughes can be cited as being just as sensible as earlier artists to the need to interpret for today's generation the meanings that can be invested in the visual scene. His collaboration with the photographer Fay Godwin in books (5) that combine poetry and landscape photography is a very good example. His interest in landscape is in a long tradition of poetics linking writers well known to every reader of English literature, such as S. T. Coleridge and W. H. Auden, to those less well known outside our shores like John Clare and George Crabbe. A scan of the poetry shelf of any good bookshop will show too that today's poets are still looking at and interpreting our native landscape.

So we could go on; but the point is made that here is a powerful similarity between Japanese culture and the UK's brand of European culture, both in historic and (I would guess) present-day terms. For many, the 19th century art critic and architectural historian John Ruskin summed it up when saying that 'fine art is that in which the hand, the head and the heart of man go together'. Japanese readers might like to ponder that, to see if they can assent to it not only for fine art (meaning painting in this case) but for the whole environment and indeed perhaps for total living.

Humans and nature in Japan

As Watanabe Masao has pointed out (5), in Japan nature was a unity and man lived in this as part of this unity. Nature was not, as in the west, something of a lower order and therefore to be challenged, altered and conquered. He cites historical examples of monks waiting until it is light enough to see the insects in their path before going out in the morning, and of the monk Ryokan leaving a leg outside the mosquito net during the night so that they too might have their supper. Watanabe suggests that this type of view is still not totally extinct in Japan but on the other hand it has not been totally replaced by western views, thus leaving the Japanese without any strong guidance at the moment.

A complex analysis of one aspect of Japanese culture is presented by Augustin Berque (6) when he examines the idea of *fudosei*, which he translates as 'mediance' as a kind of shorthand term for a concept such as the structural occasion of human existence. He argues for the non-dualism of man and nature through the ambivalence of *fudo* which is a relation which partakes of both the subjective and the objective; *fudosei* is at once the ecological character and the symbolical meaning to humans of a given milieu. There is therefore an insistence on pure experience which places the relation between a subject and its world as the first reality. So the world is not there to be commanded; it is there as a set of relationships which bear the weight of constant adaptation. The subject does not occupy a definite place according to which everything else has to be ordered: reality is to be found also between nature and culture, between ecology and symbol.

From our deep roots and a common type of history it might be possible to distill the best of the Western ideas of progress which in Britain are allied to a reverence for the past (too much so for some), to the European idea of the pre-eminence of the life-world of the individual which can give value in a time so much dominated by the needs of the crowd and the noisy majority. If we could add to these the non-duality which Japan has cherished down the ages and the way in which the harmony of the group is to be prized above the possession of dominance by the individual, then we might effect a real change in the world and its institutions. For in order to achieve a better *modus vivendi* between the rest of the globe and its ecological dominant many deeply-held ideas of class, gender relations and hierarchy will also have to be changed.

In the East, Lao Tzu said, 'to know that you have enough is to be rich' and in the west, Thoreau said more or less the same with 'A man is rich in proportion to the number of things he can afford to let alone'. With these mottoes, it ought to be possible to find common ground from which to start and a common endeavour to carry us forward. A glass of good whisky is always an excellent start.

Note
I am grateful to my friends in Japan and the UK for their interest in my work, notably Douglas Pocock, Ito Koji, Tawara Hiromi and Kuroasaka Miwako. An extended version of this essay is to appear (in Japanese) in M. Kurosaka (ed), 1990, *Resonance in Nature, vol 3 Towards Creative Relations with Nature* , Tokyo, Shikusha Publishers Inc.

Notes and references

1.　Both nations have had periods of xenophobia: that of Tokugawa Japan is well-known. Earlier this century, the headline of a London newspaper is said to have read, "Fog in Channel. Continent Isolated".

2.　A recent compilation of value on ecological ideas is R.P. McIntosh, 1985, *The Background of Ecology: Concept and Theory*, Cambridge University Press. See also J. Worster, 1977, *Nature's Economy. A History of Ecological Ideas*, Cambridge University Press; and M. Sagoff, 1985, 'Fact and value in ecological science', *Environmental Ethics* , 7, pp.99-116. A minor piece of human-historical ecology is I.G. Simmons, 1989, *Changing the Face of the Earth: Culture, Environment, History*, Oxford: Blackwells.

3.　N. Evernden, 1985, *The Natural Alien*, Buffalo & Toronto:University of Toronto Press. A useful beginning for the study of Heidegger in the present context is M.F. Zimmerman, 1983, 'Toward a Heideggerian ethos for radical environmentalism' *Environmental Ethics*, 5, pp.99-131; L. Westra , 1985, 'Let it be: Heidegger and future generations', *Environmental Ethics* , 7, pp.341-50. In English, try M. Heidegger, 1977, *The Question Concerning Technology and Other Essays* (trans. W. Lovitt), New York: Harper & Row,

4.　T. Hughes and F. Godwin, 1979, *Remains of Elmet: A Pennine Sequence*, London & Boston: Faber and Faber. An interesting examination of the context of landscape in terms of text and poetics is J. Turner, 1979, *The Politics of Landscape*, Oxford: Basil Blackwell.

5.　M. Watanabe, 1974, 'The conception of nature in Japanese culture', *Science*, 183, pp.279-282.

6.　My source for this is A. Berque, 1987, 'Some traits of Japanese *fudosei*", *The Japan Foundation Newsletter*, XIV, no 5, pp.1-7. This short article refers us to his book *Le Sauvage et l'Artifice: les Japonais Devant la Nature*, Paris: Gallimard, 1986.

Towards a Macro-Spatial Interpretation of Operation Flood

P. J. Atkins
University of Durham

Operation Flood (OF) is a major dairy development project whose aim is to provide sufficient cheap, pure milk to match the hitherto frustrated latent demand in India. Dairy products are highly valued by Indians but in the recent past supplies have been restricted and retail prices high. The purpose of this short paper is to outline some geographical insights on the structure of this scheme. Further discussion of OF has been published elsewhere (Atkins 1988, 1989a, 1989b).

Milk production

There is no specialist dairy region in south Asia and, although milk is the second most valuable output of the sub-continent's agriculture, it has hitherto been essentially a by-product of arable farming. Bullocks and buffaloes are the main energy source for ploughing and female bovines have been valued mainly for their male offspring. Traditionally most of their milk has been for calf rearing and any surplus consumed on the farm. Livestock are fed on crop residues, weeds and wayside grasses which are of low value and would otherwise be waste, and as a result the yields of India's 250 million bovines are very low. Until the last decade or so most urban supplies were procured from close to the markets by small-scale, informal sector middlemen. Supplies were inadequate, especially in the summer lean season, and quality was often poor due to adulteration and souring. The inefficiency of this system was evident in the 1980s when shortages led to rationing of milk in some metropolitan centres.

In 1970 OF was conceived with the aim of meeting the shortfall by importing skimmed milk powder (SMP) and butter oil (BO). The idea was to reconstitute the SMP, sell the milk to urban consumers, and then use the counterpart funds for building up a rural supply infrastructure of primary village co-operative societies, processing plant and transport facilities. OF has fallen into three historical phases:

- OFI (1970-81) concentrated on Bombay, Delhi, Calcutta, Madras and eighteen of the milksheds with most promising potential for development.

- OFII (1978-85) was to cover 160 cities of over one million population, and ten million rural producers.

- OFIII (1985-94) is a planned expansion to 450 towns and cities, 50,000 dairy societies, 176 milksheds, and an average procurement of 13 million litres per day.

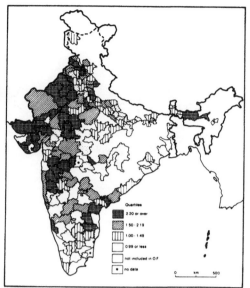

Fig. 1

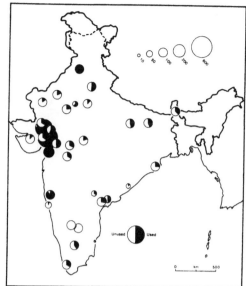

Fig. 2

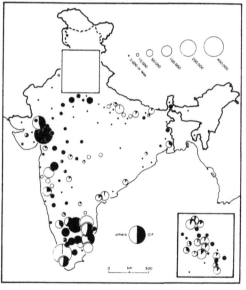

Fig. 3

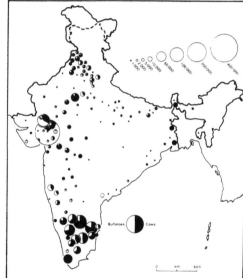

Fig. 4

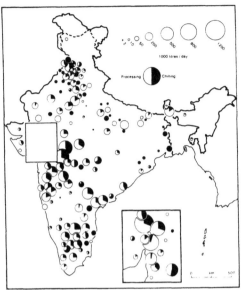

Fig. 5

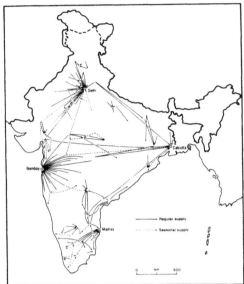

Fig. 6

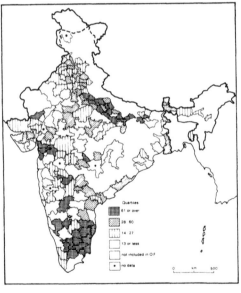

Fig. 7

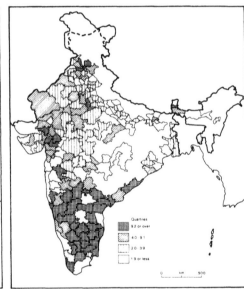

Fig. 8

152

Some of OF's targets were over-ambitious from the outset or have not been achieved because of local political or organizational difficulties. Results so far are partial, but nevertheless impressive. The following maps seek to illustrate selected geographical aspects of OF in the mid-1980s.

Figure 1 is a map of the mean daily procurement per village dairy co-operative member in kilogrammes, averaged over the year 1985/6, intended to illustrate OF's spatial selectivity. As anticipated from the regional distribution of India's milch animals, OF's supply is highest in the milksheds of Gujarat, Rajasthan, Maharashtra and coastal Andhra. It is surprisingly low, however, in the southern cone of the country. One reason for this that in Tamil Nadu large numbers of poor farmers and landless labourers, without any real resources for milk production, join their local co-operative in the hope of betterment. Apparently their certified society membership is one necessary qualification for acquiring credit to buy a milk animal, but many remain dormant or marginal members and contribute little or no milk. It would be interesting to see data on active/dormant membership of dairy co-operative societies (DCS) in the various milksheds.

This rural production is supported at the District Union level by the supply of various services. II *Figure 2* we see the location of cattle feed factories which have been built to provide balanced, compound feed to supplement the milch beast's traditional diet and boost yields. Gujarat has done well but other parts of the country have as yet seen less investment. There seems to be a problem of the underuse of existing facilities, although conclusions about overcapacity would be premature since several factories have only just started production.

Figure 3 is a record of the artificial insemination programme in India. OF provides a high proportion of AI in some districts especially, as *Figure 4* shows, for cows. The relative neglect of buffaloes in states other than Gujarat is a source of concern. A growing dependence upon the imported genes of high yielding cattle from other parts of the world may lead to problems similar to those experienced in the early phase of the Green Revolution in grain. Heat stress is possible in animals with one third of their blood from temperate latitudes, and the offspring bullocks are less suitable for ploughing than their native cousins. Cows, whose milk is not as rich as that of the buffalo, are expensive to keep because they need heavy feeding. Foreign cattle are also more susceptible to diseases endemic in India and therefore expose the farmer to an unknown risk.

Processing and transport

Figure 5 shows chilling and processing capacity provided by District Unions and State Federations. The construction of these facilities has been a priority of Operation Flood and four regions seem to have benefited most: the north west cluster from Delhi to the Punjab, Gujarat, Maharashtra and Tamil Nadu. A fierce debate has raged about the appropriateness of the high technology employed in these plants, the foreign machinery installed and the input of overseas consultant expertise.

India's remarkable array of ecological and socio-economic niches makes the planning of any nationwide scheme like OF an uncertain process. Inevitably there will be some regions with surplus milk and others with a deficit, but linking them into a trading network requires the sort of physical infrastructure and organizational skills which are lacking in most developing countries. India's National Milk Grid *(Figure 8)* seeks to overcome the problem of establishing a balancing mechanism for perishable commodities like liquid milk by providing intra- and inter-regional connections of mutual support. Both road and rail transport are involved, focusing supplies initially on the larger cities, but now spreading down the urban hierarchy. The most spectacular achievement is the regular tanker train plying the 2,000km from Anand to Calcutta.

Social context

Critics of OF accuse it of neglecting the poor. They see the scheme as an opportunity for already prosperous rural producers to enhance their wealth and power. The available evidence for testing this hypothesis is sparse, but we do at least know that the participation rate of small and marginal farmers and landless agricultural labourers in dairy co-operatives is approximately in proportion to their representation in the wider rural community. This suggests that, although OF is not specifically setting out to target the rural poor, it is at least attracting an unbiased cross-section of producers, more than can be said of other, similar large-scale development programmes. *Figure 7* records the percentage of Dairy Co-operative Societies in each district comprising scheduled castes, scheduled tribes, and other backward castes. The proportion is high in Tamil Nadu, for reasons explained above, and also in a southern strip of Uttar Pradesh and the tribal districts of south Gujarat. Performance in other parts of the country looks disappointing, but really one needs a further map showing the fraction in each area who have joined: only then can the actual and potential figures be compared.

Figure 8 plots the percentage of female members enrolled in DCS. The result is a depressingly polarised map, with women in the north apparently allowed to play little part in OF, and their sisters in the south in only a marginally better position. This reflects the well-known underlying cultural realities of Indian society. OF has a moral duty to enfranchise women economically wherever possible.

Conclusion

OF is a mirror of the ideology of Indian modernization: it has many positive aspects and some serious weaknesses that must be addressed. In the final analysis it must be remembered that the experience of Gujarat is unique, with its mix of favourable physical circumstances and spontaneous growth. To expect the transplanted Anand 'model' to take root in other, dissimilar contexts is geographically naive and one anticipates the need for a greater flexibility of planning structure if OF is to have a truly nationwide scope. The OF authorities have yet to establish a Geographical Information System which could provide the basis for a sound spatial strategy and the assessment of spatial outcomes. There is an urgent need for the type of database that would allow both daily monitoring of the National Milk Grid and the longer term modelling of the changing spatial structure of OF.

References

Atkins, P.J. (1988) 'India's dairy development and Operation Flood', *Food Policy* 13, 305-12

Atkins, P.J. (1989a) 'Operation Flood: dairy development in India', *Geography* 74, 259-82

Atkins, P.J. (1989b) *The geographical structure of Operation Flood*, unpublished ms

Figure captions

1. Mean daily procurement (kg) per dairy co-operative member, by districts (1985/6)
2. Cattle feed manufacturing capacity, by districts (tonnes per day), and its utilisation (November 1986)
3. Artificial insemination under Operation Flood and other schemes, doses per district (January-November 1986)
4. Artificial insemination of cows and buffaloes under Operation Flood, doses per district (January-November 1986)
5. Chilling and processing capacity 1986, by districts (thousand litres per day)
6. Operation Flood's National Milk Grid

7. Percentage of scheduled castes, scheduled tribes and other backward castes in the membership of dairy co-operative societies, by districts (October 1986)

8. Percentage female membership of dairy co-operative societies, by districts (October, 1986)

Section 3: The Field Excursions

London Docklands Redevelopment

David Hilling
Royal Holloway and Bedford New College
University of London

The overall traffic of the Port of London at around 50 million tonnes has not changed significantly over the last 30 years but the manner in which that cargo is handled has changed dramatically. In the early 1960s most of the general cargo was in break-bulk form and handling was labour intensive, of low productivity with slow ship turn-round time and consequently only small ships could be used. Land use in the docks was very intensive. With the unitisation and especially the containerisation of cargo, the handling became mechanised and capital intensive, productivity was greatly increased and ship turn-round speeded up - larger ships could be used and far more land was required with a premium on high capacity transport links.

Port facilities were originally undifferentiated and widely scattered and in London during the 19th century a number of enclosed dock systems were built to provide facilities for rapidly expanding trade and also a level of security not possible at open riverside wharves. These docks were provided with lock gates to allow ships to float at all states of the tide, the Thames having a particularly high tidal range. In their location accessibility and general design, these 19th century docks were unsuited to the new technology of the 1960s, and between 1963 and 1983, traffic was withdrawn from St. Katherine's, London, Surrey Commercial, India-Milwall and Royal dock systems (see map) and concentrated at a small number of high-capacity terminals at Tilbury, where land was available and road access more favourable. The loss of jobs in the docks, the decline of associated industry and a decaying urban fabric made the docklands a special example of inner-city decline.

The tour started at **St. Katherine's Dock**. Completed in 1828, the dock has a contorted outline which maximises quayage in relation to land take-up, and a small lock entrance. It was surrounded by high-rise warehouses. On closure in the mid-1960s the dock was rapidly redeveloped to provide a marina, hotel, extensive office accommodation (including London's World Trade Centre and a commodity exchange) and residential accommodation. The central Ivory House, a name suggesting its former use, has been converted to residential and commercial use. St. Katherine's Dock has effectively become integrated into the adjacent Tower Bridge/Tower of London tourist zone.

158

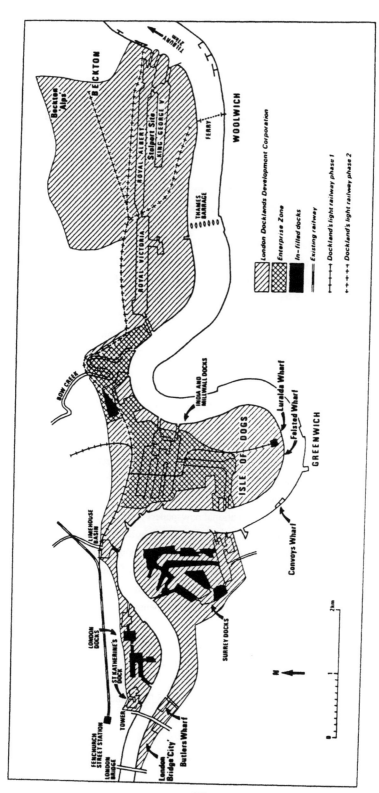

Further away from the centre of London and also on the south side of the river the pace of redevelopment was less rapid. There was a fundamental conflict between the property developers seeking high income generating activities and the local authorities favouring a mix of lower income housing and employment to replace jobs lost in the docks. In 1981 the Government created the London Docklands Development Corporation (LDDC) with powers to override the local authority planning procedures. This resulted in rapid but largely uncoordinated development over large areas of dockland.

St. Katherine's Way leads eastward into **Wapping High Street** and follows the river downstream from St. Katherine's Dock. Over centuries this was at the heart of London's port and comprised a mix of warehouses, ship servicing facilities, taverns and housing associated with the port. It is now in various stages of redevelopment. There is still evidence, rapidly disappearing, of former warehouses, redundant land devoted to temporary uses (scrap yards, storage) and residual features from the past such as taverns (the Prospect of Whitby, the Town of Ramsgate, the China Ship). Many of the multi-storey warehouses fronting the river have been redeveloped mainly for high-income housing but also for office accommodation. There are also several areas of lower income housing provided by housing associations.

The Millwall peninsula or **Isle of Dogs** provides Europe's largest urban redevelopment project. The India and Millwall Docks finally closed to traffic in 1980 and following the creation of the LDDC the developers moved in rapidly. Part of the area has been designated as an Enterprise Zone and industrial and commercial activities have been attracted by minimal planning restrictions and financial incentives. The original London fish market, Billingsgate, was relocated to the northern end of the Enterprise Zone from its original site just upstream from the Tower of London. Some of the national newspapers have relocated from Fleet Street, a large sports/entertainment arena has been built, numerous new housing developments have been located mainly on the riverside margins of the area (e.g. Cascades on the western side) and there is a variety of commercial and industrial enterprises. Some existing housing has been upgraded and a large retail hypermarket (ASDA) has been built on the eastern side of the peninsula.

The most spectacular single scheme is that at **Canary Wharf** at the northern end of the peninsula where a first phase 12 million square feet (1.1 million m^3) of office accommodation will eventually be expanded to 25 million square feet (2.3 million m^3) and includes a 50-storey office block that will be London's tallest.

The physical redevelopment is undoubtedly impressive but the whole lacks a grand design and will never be more than a number of separate sites developed in an ad hoc way. Overall, the development can also be criticised for failing to address the existing unemployment, social and transport problems of the area.

New jobs are certainly being created in large numbers but they are largely unsuited to the local population and so far have not done much more than offset continuing job losses in existing enterprises that are being forced to close. The housing is mainly too expensive for the existing population and the anticipated 50,000 employees of the Canary Wharf scheme will place impossible demands on the transport infrastructure. The £70 million **Dockland Light Railway** (DLR) links Tower Hill with Island Gardens at the southern end of the Isle of Dogs with an extension northwards to Stratford. However, it was built to minimal capacity standards and is already being upgraded by increasing station lengths to allow the use of longer trains. It is also being extended westwards into the City of London. The DLR is not adequate for existing demand and will certainly be unable to accommodate future demand.

As a tourist attraction the DLR is an undoubted success and from Island Garden it is but a short walk through a tunnel under the Thames to **Greenwich** with its variety of tourist interest. As a solution to the redeveloping area's transport problems the DLR cannot be counted successful.

The **Royal Docks** were built between 1855 and 1921 and provided London's most extensive dock system. They were closed between 1974 and 1982 and provided a vast area for redevelopment but so far progress has been slow. This probably reflects the greater distance from the City, the inadequacy of transport links and concentration of effort on the front-running Canary Wharf project. Rising interest rates have brought a general slow down in development activity. The lesson of the Isle of Dogs must also be that more time and effort must be devoted to preliminary planning and the overall grand design for the area.

Already in operation is the **London City Airport** developed by the Mowlem Group on the former quay area between the Royal Albert and the King George V docks. The runway (762m) provides only for aircraft with a short take off and landing (STOL) capability and 50-seater Dash-7 aircraft link London with Paris, Brussels and Amsterdam. The attractive new terminal is under-utilised and the airport has failed to reach its traffic targets. The operators are seeking permission to operate jet aircraft which would allow a greater range of services and attract more passengers.

The airport undoubtedly suffers from inadequate access. A proposed extension of the DLR to Beckton will help but will not eliminate the problem and a river bus service from central London to a new pier at the western of the docks still leaves passengers some distance from the airport terminal. The large water area of the Royal Docks provides great potential for water based recreational activity that could well be incorporated into the longer term planning proposals for the area.

The tour attempted to demonstrate the reasons for and the character, scale and problems associated with dockland redevelopment in London. It also served to illustrate the role of waterfront revitalisation in the wider context of inner-city regeneration.

From Steel to Sport: Structural Change in the Sheffield Economy

H. D. Watts
University of Sheffield

During the last decade Sheffield lost half its manufacturing employment but the manufacturing sector still employs about 60,000 workers or one-quarter of the city's employed population. The industrial landscape of Sheffield provides striking evidence of both surviving industries and the re-use of former industrial sites. The manufacturing activities of this city of over half-a-million people are located mainly along the valleys of the River Don and the River Sheaf. The activities in the River Don Valley are the most important and a distinction can be drawn between the Upper Don (north-west of the city centre) and the Lower Don (north-east of the city centre). The field excursion was confined to the Don valley industrial area and was designed to show participants the key features of the contemporary manufacturing economy and the new land uses emerging on former industrial sites.

Sheffield has a highly specialised manufacturing structure, two industries accounting for 50 per cent of the manufacturing employment and four industries account for 70 per cent. There are virtually no plants associated with new technologies such as data processing and instrument engineering and the two dominant industries are metal goods and mechanical engineering. The two industries of secondary importance are food and drink and metal manufacture.

The Upper Don valley has a significant representation of the food and drink industries. Along the valley floor are three important breweries, two bakeries, and one of Sheffield's largest factories with over 2000 employees. Bassett's factory manufactures sugar confectionery and is known for the production of liquorice allsorts. These industries have seen less employment contraction than the other sectors but one major food plant, producing canned foods and once employing nearly 1000 workers, has been shut down. Most of its work was transferred to a modern factory in a smaller town with a less militant work-force. The factory has been purchased by a property company and, in an unusual development for Sheffield, it has been sub-divided into units to be rented to small firms. For most older Sheffield factories conversions of this kind are not possible because of the specialised buildings constructed for metal based activities. Demolition not conversion is the norm on Sheffield's former industrial sites.

Examination of the Lower Don valley began with dramatic views over the metal based complex from a vantage point north of the city centre. Here, in the Lower Don and adjacent industrial areas are four of Sheffield's major metals sector plants. These include Sheffield Forgemasters (metal goods) on the valley floor, Tinsley Wire (metal goods) on the valley floor and the lower slopes of the southern side of the valley, British Steel (stainless steel metal manufacture) also on the south side, and Davy-McKee (mechanical engineering) 1km to the east, manufacturing industrial plant and machinery. It is in this Lower Don valley that the now closed major steel plants were situated.

From the vantage point, the Atlas Steel Melting site of Johnson Firth Brown (JFB) shows the results of demolition with no redevelopment and the building foundations are still visible. The site became redundant when JFB steel operations were merged with those of Sheffield Forgemasters and steel making was retained at the site furthest from the city centre. The Atlas site lies opposite the headquarters of Sheffield Development Corporate (SDC), a body funded by central government to lever funds out of the private sector in order to encourage regeneration of part of the Lower Don valley. The site is now owned by SDC who plan to use it for a show-piece development.

A second major steel plant was formerly operated at Meadowhall. This was shut down when its multinational owners (Lonhro) decided to cease steel making activities. This highly visible site adjacent to important north-south motorway and rail links is being replaced by a major out-of-centre regional retail and leisure development. It has over 10,000 parking places and its own rail and bus stations. Most leading UK stores (such as Marks and Spencer, Debenhams, House of Fraser, Sainsbury) occupy major sites along the enclosed shopping mall. Overall, there will be 116,000m^2 of shopping space when the development opens in September 1990. The leisure component of the site has yet to be decided.

A short distance away lies the site of Tinsley Park steel works, once part of British Steel. One of three engineering steel plants in the Sheffield area (the others are at Stocksbridge - to the north - and at Rotherham - to the east) it was selected for closure in the face of recession since, of the three plants, it had few ancillary steel related activities linked to it. Built as a new plant in the 1960s, it had a life of under thirty years. This site is now the location of a major open-cast coal working developed by British Coal. Although city planners were initially unhappy with the idea of a major extractive industry within the city boundary, their opposition was overcome when British Coal offered to include a runway for a city airport in the reclamation plan to be implemented once the coal had been extracted.

Equally dramatic changes were evident at the former site of another major steelworks. Once employing over 1,000 nothing now remains of a plant shut down by Lonhro in favour of the more peripheral Meadowhall site (now also closed - see above). On this site is emerging a sports stadium with seating for 40,000 and encompassing an Olympic standard 400 metre eight-lane athletic track. On an adjacent site, it is intended to construct a 10-15,000 seat arena. Construction of the arena has not started at the time of writing. These are some of the several sports related facilities being constructed to provide venues for the World Student Games (Summer Universiade) to be held in Sheffield in 1991. They will then make-up the framework of a major 'recreation' based initiative to ensure longer term spin-offs associated with Sheffield as a city of sport. Participants in the excursion recalled the striking artificial ski-slope built on the side of the Upper Don creating an Alpine scene of ski-lifts and skiers less than a kilometre from gasometers and derelict industrial land.

The overall image of Sheffield in the 1990s is of a city with a strong manufacturing sector accompanied by dramatic additions to the service sector. It is hoped these new service sector developments, like manufacturing, will draw income from regional, national and international sources and stimulate the development and expansion of the local economy.

Pressures in and on the Peak District National Park

Rod Brown
University of Sheffield

Note: In the following paper, reference numbers - e.g. SK 160 824 - are six-figure grid references relating to the British National Grid of the Ordnance Survey.

Purpose of Excursion

The purposes of the excursion were fourfold, namely

1. To introduce our Japanese guests to the Peak District National Park.
2. To demonstrate the major contrasts in physical environment within the National Park.
3. To illustrate the variety of land uses.
4. To discuss the environmental impact of competing and conflicting land use activities.

The Peak District National Park

Designated in 1951, this was the first of 10 National Parks established in the 1950s. These National Parks arose from a growing public demand for access to, and recreation in, areas of countryside of outstanding natural beauty. At the same time there was concern for the protection of such areas from the effects of industrial and residential development.

The Peak Park comprises 1,400 sq. km at the southern tip of the Pennines, an area of very varied landscapes, including gritstone moorlands, lush limestone dales, rugged crags and pleasant river valleys. This attractive scenery lies so close to the major conurbations of central and northern England that it is no surprise that the Peak Park is the busiest of our National Parks with some 20 million visitors in 1986. About 16 million people live within two hours drive of the Peak Park and as a consequence some 90 per cent of visitors come just for day trips. This pattern of usage creates congestion in villages and on highways, especially during Public Holidays, yet provides little income for the 40,000 inhabitants of the area.

Although some 70 per cent of the Peak Park is in private ownership the Peak Park Joint Planning Board (PPJPB) is the independent public authority with ultimate responsibility for planning policy. However, their policies and decisions must be in accordance with the National Parks Act of Parliament 1949 (and later amendments), which define the following objectives for all our National Parks:-

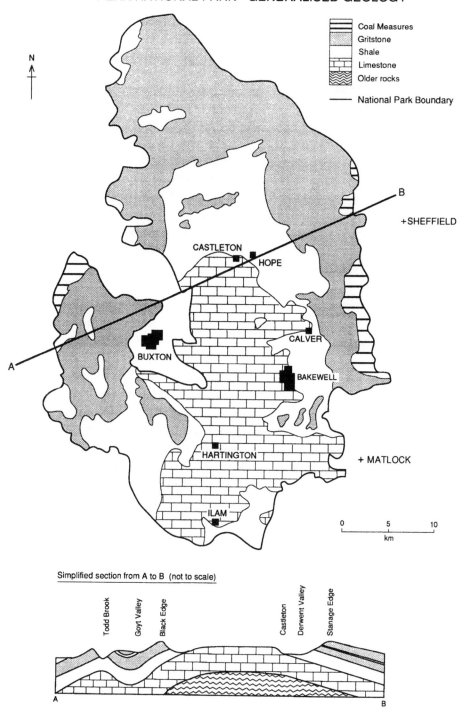

PEAK NATIONAL PARK - GENERALISED GEOLOGY

Coal Measures
Gritstone
Shale
Limestone
Older rocks

National Park Boundary

N

B

+SHEFFIELD

CASTLETON
HOPE

CALVER

BUXTON

BAKEWELL

HARTINGTON

+ MATLOCK

A

ILAM

0   5   10
km

Simplified section from A to B (not to scale)

Todd Brook
Goyt Valley
Black Edge
Castleton
Derwent Valley
Stanage Edge

A

B

1.   to preserve and enhance natural beauty
2.   to promote open air recreation for the public
3.   to preserve wildlife and historic buildings
4.   to maintain viable local communities

Since 1951 there have been many cases where objectives (1) and (3) come into conflict with (2) and (4). Some of these cases will be considered during the field excursion.

Although the physical landscapes of the Peak Park are a product of many environmental influences, there is no doubt that geology is a key variable influencing the scenery we perceive today. The names White Peak and Dark Peak are used to distinguish two strikingly different areas of the Peak Park. The White Peak being a limestone area with a network of white walls and frequent pale rock outcrops. In contrast, the Dark Peak is an area of gritstone and shale, its dark appearance resulting from moorland vegetation, bare peat exposures and gritstone crags.

General transect

The excursion followed minor roads through the western suburbs of Sheffield to a short stop at the National Park boundary at **Ringinglow** (SK 291 837). The large millstone on a plinth being the emblem of the Peak Park. A brief stop at **Upper Burbage Bridge** (SK 261 830) gave fine views of Dark Peak moorlands with peaty soils clad in heather and cotton grass. The first main stop was a walk from **Stanage Edge** (SK 252 827) to **Stanage Plantation** (SK 238 837), (see *Locality 1* description). From *Locality 1* minor roads were taken past **Bamford Edge** (SK 210 847) to join the A6013 leading north to **Ladybower** (*Locality 2*). From Ladybower a minor road through Thornhill led onto the A625 and then through Hope village to the car park at **Castleton** (*Locality 3*). From Castleton a minor road led to a viewpoint (SK 158 825) overlooking **Pindale** (*Locality 4*) and the **Hope Cement works** (*Locality 5*). The return journey to Sheffield followed the A625 through Hathersage village (SK 235 815).

Locality descriptions
Stanage Edge (SK 252 827 to SK 238 837)

The party walked for 3km along the magnificent escarpment of Stanage Edge. The caprock is the Chatsworth Grit, a 50m thick sandstone (gritstone) bed within the Millstone Grit Series, a sequence of alternating sandstones and shales of Carboniferous age. At Stanage the Chatsworth Grit stratum dips at two to three degrees to the east and so the dip slope is a gentle incline towards Sheffield whilst the craggy scarp faces west and overlooks the Derwent Valley. Beneath point 457m (SK 251 830) circular millstones were examined on blocky slopes below long abandoned quarries. The millstones were carved from blocks of Chatsworth Grit, here a coarse textured feldspathic sandstone containing quartz pebbles up to 30mm in diameter. Stanage millstones were used mainly for wood pulping though some of the finer textured stones were used for tool and cutlery grinding in Sheffield. Due to falling demand and imports of better quality stones millstone quarrying ended in the late 19th century.

The walk north westwards along the crest afforded excellent views of crags which are important rock climbing resources. With over 650 climbs of up to 35m in height the popularity of Stanage is such that on a Sunday in summer over 1,000 climbers may be on the edge. The descent through Stanage Plantation (238 837) traversed bouldery slopes of solifluction deposits. These were formed during periglacial conditions in the late Pleistocene.

The area visited is part of the North Lees Estate, formerly a private estate bought in 1971 by the Peak Park Planning Board. The estate is largely sheep farming land comprising moorland, rough grazing and improved pasture. The land is also increasingly used for outdoor recreation, including walking, climbing and hang gliding. Agriculture and recreation are competing and conflicting land uses and the Peak Park Planning Board faces difficult compromises in managing this land. Problems created by recreation include footpath erosion, moorland fires, litter, and the disturbance of sheep and lambs by dogs.

Ladybower Reservoir (SK 202 856)

With a capacity of 2.86^{10} litres Ladybower is the largest of the Upper Derwent reservoirs. Completed in 1945 it was the last South Pennine reservoir built to cater for the growing demand for water in the surrounding urban and industrial areas of Manchester, Sheffield and the Midlands. The catchment area rises to over 600m and receives about 1,500mm rainfall a year. Being largely moorland with a land use of sheep grazing, grouse rearing and outdoor recreation this area is ideally suited to water catchment. The deep and confined Derwent valley afforded a good site for the construction of an earthfill dam. Nevertheless, deep excavation and grouting of the dam foundations were required to ensure that leakage did not occur.

CASTLETON - LOSEHILL HALL - PIN DALE - HOPE CEMENT WORKS
LOCATION AND TOPOGRAPHY

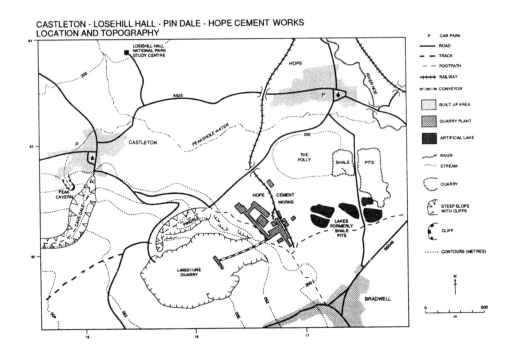

LOSEHILL HALL - PIN DALE - HOPE CEMENT WORKS
SOLID GEOLOGY

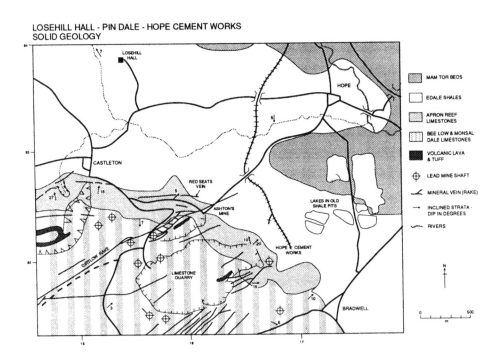

There was opposition to the building of the reservoir both from environmental conservation groups and also from the inhabitants of the villages of Ashopton and Derwent which were to be submerged. The displaced village people were rehoused at Yorkshire Bridge (SK 201 850), just below the dam. Reservoirs certainly add diversity, and some would say beauty, to the scenery of the Pennine valleys; but there are many who feel that they are an unacceptable alteration to this landscape.

Castleton (SK 150 830)

In 1080 the Norman nobleman William Peveril built a castle on the limestone knoll overlooking Peak Cavern. The Normans used the castle as a lodge during hunting forays into the woodlands of Peak Forest. About 100 years later Castleton was founded and soon it became a thriving village. Decline set in during the 14th century and by the 17th century the castle was in ruins and Peak Forest much denuded. Castleton did not prosper during the lead mining boom of the 18th century for it was peripheral to the main centres of production.

With the growth of motoring in the 1920s an increasing number of visitors were discovering the attractions of Castleton. This trend accelerated after the designation of the National Park in 1951 and the village is now the main centre for tourists visiting the north eastern part of the Peak Park. The reasons for this popularity include:

    (i)   an impressive site below craggy limestone slopes,

    (ii)   spectacular dry valleys of Winnats and Cave Dale,

    (iii) show caves at Peak Cavern, Speedwell, Treak Cliff and Blue John,

    (iv) the ruins of a Norman castle,

    (v)   rugged terrain of the landslip at Mam Tor (SK 130 837),

    (vi) many facilities for visitors, including shops, hotels, cafes and car parking.

Pin Dale (SK 160 824)

This dry valley, much altered by limestone quarrying and lead mining, was one of many sources of lead ore in the 18th and 19th centuries. The ore was found in a mineral vein (Dirtlow Rake) on the north side of the valley.

Archaeological evidence suggests that Peak District lead was first worked by the Romans. After reaching a peak in the 18th century mining rapidly declined as demand fell and imports became cheaper. Some lead is still produced as a by-product of fluorspar mining, an activity which is second only to limestone quarrying in economic value in the National Park.

The Hope Cement works (SK 168 823)

Because the works lies at the geological boundary between Carboniferous Limestone and Edale Shale the two key raw materials, limestone and shale, are readily available. There are shale quarries north of the plant and a vast limestone quarry to the south. The works produces about 1.25m tonnes per annum of cement, some eight per cent of the U.K. output, and employs over 500 local people. Ever since the works was established in 1933 there has been a conflict of opinion between those wishing to conserve the environment and those wishing to exploit the mineral resources. This conflict became more sharply focussed after the National Park was designated in 1951. Since then there have been several public inquiries into the future of the plant and associated quarries. The long term future of the works is uncertain because recent applications by the company to extend the limestone quarry have been refused. At present extraction rates the available limestone reserves will run out about 2035, whereupon the works may no longer be viable. No other single industrial enterprise in the National Park has aroused so much public opposition on environmental grounds as the Hope Cement Works.

References
i) Ordnance Survey Maps
1:63,360 *Peak District Tourist Map 4*
1:25,000 *Outdoor Leisure Map 1*, The Dark Peak
1:25,000 *Outdoor Leisure Map 24*, The White Peak

ii) British Geological Survey Maps and Memoirs
1:50,000 *Sheet 99* Chapel en le Frith (1983)
1:50,000 *Sheet 111* Buxton (1978)

iii) Books and Pamphlets
Anderson, P. and Shimwell, D., 1981, *Wild flowers and other plants of the Peak District.*, Moorland Publishing Company.
Briggs, D.J., Gilbertson, D.D. and Jenkinson, R.D.S., 1985, *Peak District & Northern Dukeries,* Quaternary Research Association.
Edwards, K.C., 1962, *The Peak District,* Collins.
Millward, R. and
 Robinson, A., 1975, *The Peak District.*, Eyre Methuen.

iv) Peak Park Joint Planning Board (PPJPB) publications:
First and Last, 1979, *The Peak National Park in words and pictures* .
Ford, T.D. & Rieuwerts, J.H., 1983, *Leadmines in the Peak District.*
Peak District National Park Approved Structure Plan (1982).
A Guide to Policies affecting the Peak District National Park (1983).

Transect across the South Yorkshire Exposed and Concealed Coalfield

Bryan E. Coates

University of Sheffield

Note: In the following paper, reference numbers - e.g. 43/331940 - are six-figure grid references relating to the British National Grid of the Ordnance Survey.

Purposes of the Excursion

1. To introduce our Japanese guests to one of Europe's major coalfields.

2. To show how the development of the coalfield through time and space over the last 150 years or so has resulted in different socio-economic structures (including settlement patterns) and problems in different parts of the coalfield.

3. To illustrate the significance of the depth at which coal is worked for size of colliery operation and size (and age) of associated settlement, and hence the geography of the coalfield.

4. To recognize the significance of particular coal seams (e.g. Silkstone, Parkgate and Barnsley) in the exploitation of the energy source.

Places Visited

General transect - ran from southwest to northeast across the coalfield. Hence we were going "down dip" and at right angles to the strike, the outcrops running roughly parallel to each other in a NW/SE direction. The older seams in the carboniferous deposits are exposed in the west and the younger seams are covered by the Magnesian Limestone, Permian and later deposits in the east. It is most important to note that the Exposed Coalfield in South Yorkshire is unusually wide and therefore huge quantities of coal have been won from relatively shallow depths. Not until this century were pits sunk in the Concealed Coalfield, and all but one was sunk initially to work the relatively thick and undisturbed Barnsley Bed (i.e. the coal seam named after the town of Barnsley where it outcrops).

Particular sites/zones visited

Grenoside/Wortley (43/331940 to 43/308995) Running along the strike with Lower Coal Measures and Millstone Grit to west and Middle Coal Measures to the east. Collieries are hidden in the shale valleys between the sandstone cuestas. A very rural scene of former medieval deer parks, landed estates and early 19th century enclosure.

The Silkstone sequence (43/306997 to 44/344003)

A fascinating sequence of different mining methods employed to work the same seam (Silkstone) of good quality coal - from west to east we see medieval bell pits, recent open adit mine (now completely gone), a shaft-cum-drift mine closely associated with the booms and slumps of the industry in the last 80 years' and a deeper mine at Pilley. At both ends or the sequence we have "estate villages".

The Worsborough area (44/350027)

The estate village (church, inn, hall, park, dwellings) of the Edmunds family. Worsborough illustrates well the importance of land ownership in the development of the coalfield. The outcrop of the Barnsley Bed was particularly thick in this area and the feeder canal with its reservoir was dug to get the coal to a wider market. After nearly two centuries of dependence upon coal the area is now characterised by pit closures and the associated dereliction.

Southern side of the Dearne Valley - from Stairfoot (44/373057) to Mexborough (44/478000)

A more or less continuously built up area of terrace housing and newer estates. Stairfoot is an excellent place to reflect on the importance of transport facilities in the development of the coalfield, and above all the railways. Roads, canal and railways form a gordion knot. Today the landscape is largely devoid of the collieries that formed the settlements and associated facilities and much of the ill-drained valley floor is wasteland (at Manvers (44/452009) we saw huge piles of rotting coal cutting and conveying machinery). Occasionally the old villages - cores of the mining towns - can still be discerned on the better drained slopes above the flood plain. We had our lunch stop at the best example (Wath upon Dearne, 44/433009).

Northern side of the Dearne Valley particularly Goldthoroe and Thurnscoe East (44/462044 to 44/468054)

Settlements are discrete on the northern side of the Dearne Valley, pits are deeper and some are later and located on the eastern margin of the exposed coalfield. Brick dwellings have replaced stone but the terrace is still the initial building form. We reconstructed an imaginary walk from Goldthorpe village to Hickleton pit in the great strike of 1984. The pit is now closed after almost a century of mining coal.

Hooton Pagnell (44/486080)

This is a rural enclave without a colliery in view. With its Norman church, attractive hall, and gable end to the street farmsteads one can imagine oneself to be far away from Britain's main coalfield. The village is stretched along the top of the scarp of the Magnesian Limestone outcrop.

Brodsworth Colliery/Woodlands estate village (44/53077)

A huge colliery sunk in the 1900s. The colliery company built an unusually elaborate, spacious and attractive estate village for its employees and this helped to attract the best workers from all the coalfields of Britain. Each colliery in the east worked a large area (the "take") and, in consequence, the communities are widely spaced and separated by stretches of rural land.

Major issues/problems in the South Yorkshire Coalfield as indicated by places visited

Current socio-economic problems

The rapidity of change since the end of the 1984 strike has produced a series of interconnected social and economic problems. Unemployment is an obvious symptom. Underemployment and a limited range of employment opportunities is another less obvious symptom. A political aspect of these changes is the dominance of a Thatcher Government in London and its "neglect" of issues in a Labour heartland such as South Yorkshire. The huge Labour majorities in the Parliamentary constituencies of the coalfield worked against the development of a national concern and sensitivity to its problems. Indeed, they may have invited the central government to "cut them down to size".

As the pits have closed the area is left with a settlement pattern and infrastructure produced to mine coal. The pit "villages" (often with populations of 15,000 plus) are left as anachronistic implants in the landscape and their dependence on the major urban centres of Barnsley, Doncaster, Rotherham and, above all, Sheffield for jobs and shopping becomes a critical factor in their socio-economic well being.

Current environmental problems

Major problems of subsidence and waste disposal remain. Signs of abandonment, dereliction and neglect are all too common, and especially obtrusive in the flat lands in the east. Yet the majority of the area is still rural and some parts have changed little in recent decades, though the motor car has made all parts very accessible to commuters from the main towns.

The big power stations also dominate the eastern landscapes and trail up the Don Valley as far as Sheffield.

The extent to which the coal and steel image of the past is a deterrent to new developments is still an open issue. Some have argued that neglect and dereliction engender apathy and acceptance of the status quo. Yet the area is very accessible to the motorway network and straddles the Al from London to Edinburgh, and is nearer the former than the latter. It is also within easy reach of several Universities and Polytechnics, notably those of Sheffield.

A final comment...... The coal has *not* been worked out. Despite the long period of active mining there are very considerable reserves left in the ground. Whether or not they will be worked in the future will depend, as always, on the available technology, current and expected demand, and profitable proximity to that demand, which, in turn, is largely a function of transport facilities and costs.

An Excursion through Durham City

Douglas C.D. Pocock
University of Durham

The aim of the excursion was to give the geographical background and attempt to highlight the townscape qualities of a city which medieval writers likened to Jerusalem, which Ruskin termed one of the wonders of the world and which Pevsner, more modestly, called one of the architectural experiences of Europe. The picture was built up from a series of stops or vantage points.

The panorama from Wharton Park, just above the railway station, shows the historic city set in a bowl, 30 metres at its lowest, rising to a rim of 90 metres at Gilesgate Moor (to the east), Little High Wood and Elvet Hill (south) and the spine of the old A1 and Crossgate Moor (west). The bowl is occupied, and the rim breached, by the meandering River Wear, which exhibits two valley types. It is broadly retracing its preglacial course in unconsolidated material, e.g. Race Course and The Sands, and narrower where it has cut a new, postglacial course in the solid geology, e.g. gorges of the central meander loop around the cathedral and castle and, again, downstream of Kepier. The narrowness of the meander neck of the 'peninsula' (220m) is articulated in the cutting of Leazes Road (1967) through the Claypath spine.

The extensive deposit of unconsolidated fluvio-glacial sands and gravels has been deeply serrated to provide a very varied relief. The deep, blind re-entrant of Flass Vale, between Wharton Park and Crossgate Moor, is a striking example of the deposit's erosion; the railway requires a viaduct of eleven arches to span the mouth of the valley. Beyond, the railway illustrates the unconsolidated nature of the sediment in the choice of a deep cutting, rather than tunnel, through the western rim.

The rim determines intervisibility from the centre, and its green horizon has been partly used in the delimitation of the city's Conservation Area. The juxtaposition of contrasting relief forms emphasizes the elevated nature of the peninsula and contributes to the prominence of cathedral and castle. The twin silhouettes of the architectural climax change as the observer moves about the city. Many other buildings have both front and back exposed as a result of the rolling topography. A social hierarchy is respected among buildings, with spiritual and cultural uses providing the dominant markers.

The solid geology of the world-famous gorge may be interpreted from exposures and other features in the vicinity of Prebends' Bridge. The Lower Coal Measure Series of sandstone and shales is interspersed with very thin coal seams. The lowest and richest of the coal seams was formerly exploited from Elvet Colliery; remains of an air shaft are visible on the south wall of the gorge. The Low Main (or 'Cathedral') sandstone was used in the construction of the cathedral and castle. Some stone was quarried from the gorge itself: behind, the open area is the former 'Sacrist's Quarry'; the road above the southern lip of the gorge is 'Quarry Heads Lane'. Other extensive quarries were at Kepier. The gentle eastward dip of the varied strata gives rise to perched water tables and to associated springs, e.g. South Street, St Cuthbert's and St Oswald's Wells.

The River Banks as an amenity feature were consciously landscaped by the Dean and Chapter in the 18th century with the rise in promenading and quest for the picturesque. Paths were laid out to exploit the meander loop and produce a varied series of contrived vistas, the climax of which was the prospect from Prebend's Bridge, built and paid for by the prebends or canons of the cathedral in 1778.

As the bridge is crossed the cathedral towers rise above the Banks' foliage and the castle just protrudes its presence above the medieval bridge, which itself is a focal point concealing the subsequent course of the river and leaving a green skyline. (Also concealed, carefully positioned behind Framwellgate Bridge, is the new vehicular bridge at Millburngate). It is up to the observer to decide whether the attraction of the viewpoint lies in the components of the view or in their classical arrangement or perhaps potential for 'prospect and refuge', or in their symbolic or historic value. Certainly the view has been acclaimed by painters, writers and scholars alike: Turner, Hawthorne and Pevsner, for example, acclaim it as *the* view of Durham.

The footpath ascent out of the gorge brings one to South Street. Once part of the Great North Road, this narrow street flanks the western lip of the gorge, descending increasingly steeply to gain entry to the former walled city across the ancient Framwellgate Bridge. The individual houses lining the 'landward' side (mainly 18th to 19th century facades, often with older structures behind) constitute unexceptional architecture but, collectively, constitute attractive townscape. On the opposite side of the street is the theatre of the sublime. From here can be obtained the most complete view of the Norman citadel of castle, cathedral and abbey which Pevsner termed one of the great architectural experiences of Europe. One can also appreciate the recent achievement of inserting a university library (George Pace, 1967; Civic Trust award) between castle and cathedral.

Large buttresses at the west end of the cathedral indicate that building approached precariously close to the edge of the near-precipitous slope. (The section of the Banks is known as 'Broken Walls'). The depth of the wooded gorge is 30 metres. The western towers, revealing interesting construction details, rise to 47 metres; the central tower is 70 metres in height. The Market Place, in the shadow of the castle, is marked by the nearer, cleaner church spire (St Nicholas).

Framwellgate Bridge is sited where the river leaves the gorge for the plain section. The castle, dwarfing the domestic scale of buildings below, towers over the northern entry to the city. It was formerly supported in its role by a gate on the bridge, while a river weir strengthened the water barrier.

Millburngate was formerly an industrial area, using water from the Mill burn (now culverted under North Road) and river Wear. Visual reminders are the Bishop's Mill and former carpet factory. Prominent modern constructions are Millburngate Bridge (1967); Millburngate House (T.F. Winterburn, 1969; 300,000 sq.ft. of office space, largely for the National Savings Certificate division); Millburngate Shopping Centre (Building Design Partnership, 1976; 68,000 square feet of retail space; Civic Trust and Europa Nostra awards).

Back Silver Street shows a haphazard arrangement among early units, which has continued in a breaking-up of recent buildings. Consequently, neither height nor mass distract from the visual climax of the city; on the contrary, the area provides a complementary foil, a 'pebbly foreshore' to the peninsular acropolis.

Recent floorscaping (1978, Civic Trust award) evokes a past era (minus its inconvenience). York paving stones and setts, with central 'wheelers', have reconstituted the street surface as an obvious linear feature and provided a pervasive design element in the narrow, winding streets of the centre.

The Market Place is the civic hub. Huddled in the shadow of the castle for its former protection, it is at the meeting point of three narrow, ancient streets on their way to the cathedral from Framwellgate and Elvet bridges and along Claypath. The equestrian statue is a prominent central marker, giving coherence to the square as well as emphasising its modest size. The harmony of the buildings lining the square is set by the church and town-hall complex (both rebuilt in the middle of the last century) and continued by three prominent banks. Even the art deco elevation of a 1930's department store conforms to the dignity of the scene, the quality of which was enhanced during the 1970s by sandblasting of buildings and by pedestrianisation and floorscaping.

The former quality of enclosure was weakened when the Claypath feeder lost its continuous building line in the construction of a central bypass close to the Market Place itself. Since the new relief road crossed Claypath in an underpass, necessitating removal of property lining the old street above, the exposed gap provides a vantage point from which to appreciate both the narrowness of the meander neck and the extent of the open-heart surgery which the inner area underwent in the 1960s and early 1970s. The centre was relieved of through traffic by the connecting of the new Leazes Road with two new bridges either side of the peninsular neck, Millburngate Bridge (1967) and another Elvet Bridge (1975). Leazes Bowl multi-storey car park (W. Whitfield, 1975; R.I.B.A. commendation) is the first reinstatement covering part of the exposed scar.

Palace Green is the climax of the excursion. At the head of attractive Saddler Street the pilgrim catches his first glimpse of the cathedral, with the twin west towers framed between the buildings lining the final narrower, curving ascent of Owengate. At the top of this short ascent, confinement gives way to unbarred climax. Across the whole width of the southern end of the green is anchored the majestic vessel of the cathedral. Flanking either side of the green are buildings - formerly ecclesiastical but now university property - all showing deference in their privileged position. The castle, the former episcopal palace, closes the northern end of the green, with the keep a distinctive marker.

Apart from its incomparable setting, the cathedral as a building is of first rank in importance - aesthetically, architecturally, culturally. It is the finest early Romanesque building in the country; it is the first building in north-west Europe with stone ribbed-vaulting; its concealed flying buttresses ushered in the Gothic; it is the centre piece of the most spectacular group of buildings in England commemorating a distinctive phase in the history of our country.

The castle offers its own sequence of experiences - Norman crypt chapel and galleries, 15th century great hall and kitchens, Jacobean chapel, sumptuous furniture and furnishings of the former prelates. The former Benedictine monastery, one of the most complete in England, offers an equal variety of experiences, both internally and externally. In 1987 cathedral and castle were designated a World Heritage Site.

Causey Arch and the Industrial Revolution
in North East England

P.J. Atkins
University of Durham

Causey Arch in County Durham (grid reference NZ 205 562) is a site symbolic of the very fundamental restructuring of the economy of the north east of England which took place in the 18th and 19th centuries. The spatial evolution of the northern coalfield was a complex outcome of the interaction of a number of significant factors. Geological conditions were of course a constraint, but we must also consider the available mining and transport technology and the facilitating organization. There is room here for only a brief resume of two main points: the attempts by entrepreneurs to monopolise space, and the outward linkages of the region with other parts of an emerging national economic system.

As some of the pits close to the River Tyne were worked out in the late 17th century, it became imperative to develop sources further afield. This could only be accomplished if ready access to the river was maintained, but pack horses or carts were slow on the poor contemporary roads, and therefore expensive. The solution was purpose-built wooden railways or waggonways along which coal could be hauled with reasonable efficiency by horse power and gravity. These linked with staithes along the river bank from which coal was landed into keels, for transport downstream to waiting ocean-going ships. Waggonways were costly to construct and 'wayleaves' had to be negotiated with landowners to cross their land. In the early 18th century these expenses were increasingly shared by partnerships of colliery owners who, by pooling their resources, were able to make investments they could not have contemplated individually. The most famous such group was the Grand Alliance formed in 1726 by, amongst others, William Coteworth, Henry Liddell, George Bowes and Edward Wortley, who attempted to monopolise space on the coalfield by buying up wayleaves and paying 'dead rents' on potential colliery sites they had no immediate intention of developing.

Causey Arch was constructed by Liddell and Wortley in 1725-6 over the Causey Burn for branch lines to join the new Tanfield waggonway, which ran 13km north east to the Tyne at Dunston. This was for several decades the longest single span arch in the country and is reputedly the world's oldest surviving railway bridge. In the early 1730s up to 500 waggons per day paid tolls to cross the bridge and then ran by gravity for much of the route northwards. Horses were used to pull the empty waggons on the return journey. A replica coal waggon (built to carry 33 cwt) and length of double wooden track are displayed near a viewpoint where visitors have a spectacular view of the 100ft deep river gorge.

As the waggons left the arch on their journey to the staithes, they passed over a substantial embankment, again built as a crossing of Causey Burn. The stream itself was culverted and many thousands of tons of fill must have been carted in. Whereas Causey Arch fell into disuse after 50 years, the embankment carried the Tanfield Railway until 1962 and a main road until the present day. The Tanfield Railway surprisingly did not use iron rails until 1839 and the introduction of steam locomotives was delayed until 1881.

The theme of intra-regional spatial monopoly must be considered in conjunction with the linkages which the north east had with other regions. Inward investment and the extraction of surplus value by externally controlled manufacturing enterprise is epitomised locally by the activities of Ambrose Crowley who in 1691 located his ironworks on the lower reaches of the River Derwent to the north of Causey.

The ready availability of water power, charcoal, coal, millstone grit for grindstones, and quantities of ironstone in the coal measures encouraged metal working and the production of edge tools in this area. In 1687 a number of German swordmakers escaping religious persecution had already migrated to Shotley Bridge. Crowley, who came from a famous nail making family of Stourbridge in Worcestershire, arrived in Sunderland in 1682 with the intention of using cheap local coal and imported Swedish pig iron. His employment of skilled foreign workers proved unpopular, however, and in 1690 he moved his enterprise to Winlaton. Crowley's ironworkers produced a wide range of goods from pots and hinges to anchors and cannon. In 1703 he installed steel furnaces and supplied the metal to the swordmakers. The business prospered, expanding to Swalwell, and in 1768 it was described as 'among the greatest manufactures of the kind in Europe'. It is interesting to note that, although factory-style production was an important element of iron and steel making, here much of finished output was the responsibility of blacksmiths employed on an outwork basis. Crowley's employees were subjected to a strict regime of rules and regulations, but a school was provided for their children and by contemporary standards working conditions were good. Most of Crowley's products were exported to London, where the East India Company was a major customer, and, since the company's headquarters were at Greenwich, the profits left

Tyneside. Ambrose Crawley was knighted in 1706 and his family kept an interest in these factories until 1863.

The north east has had a peculiarly disadvantaged, one might almost say 'colonial', type of industrial economy for the last 400 years. Primary extraction, principally coal but in the nineteenth century also lead and other minerals, was closely linked to the export of domestic and other grades of fuel, with only a relatively small quantum of locally added value. Much of the manufacturing which came to be established was subject to exogenous ownership and control. These are two of the inherent weaknesses which still haunt the region's economy.

Reference
Atkins, P.J., 1990, 'Historical geography of Newcastle', in Atkins, P.J. (Ed.) *Durham Geographical Fieldwork*, Department of Geography, University of Durham, pp. 111-27

The Beamish Museum

I.G. Simmons
University of Durham

By way of relief from the sterner business of presenting and listening to papers, the conference spent an afternoon at the North of England Open Air Museum at Beamish, Co Durham. The Museum is about 30 minutes easy travel by coach from Durham and we were able to take advantage of a fine if windy day.

Beamish, as it is generally simply known in the region, is one of the early generation of open-air museums aimed at reconstructing the life-ways of a particular period. The material for this collection was garnered from all over the region under the impetus of its founding genius, Frank Atkinson, and stored in the nearby Beamish Hall until capital was available for the layout of the site and the display of the items brought together.

The theme of this Museum is industrial life in the north-east of England in the period when it was one of the great hearths of European industrialism, based on shipbuilding, coal and steel. At this site, it is the coal which is emphasised since the scale of operations needed to depict the industries would be altogether forbidding. The majority of the exhibits, therefore, date from the period between the two World Wars, with some of them dating from before that.

The museum occupies a large site across a shallow valley, and arriving at the entrance, visitors can inspect the sweep of terrain in which various of the exhibits are placed. The valley was in fact formerly partly farmland and partly a small coal mine, with the waste land of the latter having been restored to support some of the Museum buildings. Since Beamish is an open-air museum aiming at re-creating a "total environment", its material is grouped around the site in a series of theme areas. Those which attract most attention are (1) a coalmine of the shaft variety with the appropriate steam-powered winding gear in full working order; (2) a small adit mine into the hillside into which visitors may walk, having put on the proper headgear; (3) a 19th century railway station with locomotives and stock; (4) a working farm from the 1920s with the machinery of the period; and (5) a town street with a pub, residences, offices and shops which are stocked and furnished in historic fashion. These separate sites are connected by working bus and tramway and, in the high season, by steam-powered railway.

The Seminar assembled on the hillslope above the main part of the Museum and Professor Simmons explained the layout and working of the area and made some recommendations to Seminar members about how best to occupy the limited time which was available. The UK members then dispersed themselves around the Museum to make themselves available to explain to Japanese colleagues some of the niceties of items found in the Museum, and in the case of the more senior individuals to recall from their own pasts just what it was like to have a tooth drilled using a foot-powered treadle and no local anaesthetic. Japanese members of the seminar showed an especial interest in the adit mine, with most taking the chance to venture into if not the bowels of the earth then certainly beyond the oesophagus: the stores and houses of the town also resonated with the memories of the equivalent type of place in the Japan of the 1930s which was known to some of our senior overseas colleagues.

The whole context of the Museum provoked discussion at various times on the day. The finer points of why we should spend money reconstructing the past, the role of "Heritage" in the UK today were placed alongside the far more pragmatic consideration of the attractiveness of such places to tourism and their role in the regeneration of a regional economy which has been under considerable stress for the last 40 years or so. It seemed to be the general consensus that although such places were full of selective memory in that they tended to underplay the downside of life in the selected period, nevertheless they were far better than the traditional type of enclosed museum in giving visitors an introduction to life in the past: for our colleagues from Japan the past is yet another country.

For Product Safety Concerns and Information please contact our EU
representative GPSR@taylorandfrancis.com
Taylor & Francis Verlag GmbH, Kaufingerstraße 24, 80331 München, Germany

www.ingramcontent.com/pod-product-compliance
Ingram Content Group UK Ltd.
Pitfield, Milton Keynes, MK11 3LW, UK
UKHW040927180425
457613UK00010B/279